Geology and the Hawaiian Islands

Lessons in Earth Science
A Biblical View

Patrick Nurre

Geology and the Hawaiian Islands

Lessons in Earth Science
A Biblical View

Patrick Nurre

Geology and the Hawaiian Islands, Lessons in Earth Science, A Biblical View
Published by Northwest Treasures
Bothell, Washington
425-488-6848
NorthwestRockAndFossil.com
northwestexpedition@msn.com
Copyright 2017 by Patrick Nurre.
All rights reserved.

Printed in the United States of America. No part of this book may be reproduced in any manner whatsoever without written permission except in the case of brief quotations embodied in critical articles and reviews.
Cover photo: Vicki Nurre
Title page photo: Jon Rosenberg, used by permission.

Scripture quotations taken from the New American Standard Bible®,
Copyright © 1960, 1962, 1963, 1968, 1971, 1972, 1973,
1975, 1977, 1995 by The Lockman Foundation
Used by permission. (www.Lockman.org)

Contents

Introduction - How to Use This Book	4
Lesson One - The Framework for Interpreting the Geology of the Hawaiian Islands	6
Lesson Two - Radiometric Dating and the Hawaiian Islands	13
Lesson Three - The Origin of the Hawaiian Islands	28
Lesson Four - Types of Volcanoes	38
Lesson Five - Types of Eruptions	54
Lesson Six - The Rocks and Minerals of the Hawaiian Islands	61
Lesson Seven - The Volcanoes, Earthquakes, and Land Forms of the Hawaiian Islands	71
Appendix A - Final Comprehensive Exam and Answers	105
Credits	109
Index	112

Introduction

How To Use This Book

What better way to study Earth Science than to study the geology of Hawai'i? Hawai'i is more than just black rock covered in foliage. It is a fascinating world of chemistry, minerals and the formation of lava. It is a laboratory for exploring the history of the earth as recorded in Genesis chapters 1-11. So, let's dive in and enjoy this paradise called Hawai'i.

This book is meant to give the student of geology a Biblical view of Earth history. For over 200 years, civilization has been subject to a secular interpretation of the landforms, rocks and fossils. My goal in this book is to help you develop an interpretation that does not leave the God who created nature out of that interpretation. You will be given short Earth science and Earth history lessons that illustrate geological principles through the study of the Hawaiian Islands. The lessons will be reinforced through pictures, word challenges, thought questions, activities and a comprehensive test to ensure that you are developing the critical skills necessary in correctly interpreting landforms spread across the Islands. I would suggest doing a vocabulary study on the "Word Challenges" in each section, before going on to the actual lesson. Be sure to acquire a lab book, such as a spiral notebook, to record the definitions of the word challenges, to keep track of what you are learning, and especially when you have activities to complete.

This book is meant to be studied with rocks and rock-forming minerals. Although these are not absolutely necessary to completing the course work, they do provide a crucial component of study and a perspective that just cannot be acquired by simply looking at pictures. It is highly recommended that you purchase a set of samples either from Northwest Treasures or from some other source. Also, if you live in Hawai'i, take the time to drive to the various rock formations that are located on your island. Observe and take notes. The study of geology takes time and an observant eye.

Upon completion of the course work in this book, the student should have a good working knowledge of both the Earth science and the Earth history, and a practical guide in how to interpret the wonders of nature, of the Hawaiian Islands.

Recommended Materials:
- Lab book such as a spiral notebook
- A set of volcanic rocks that include basalt (as pahoehoe, *A'a'*, cinders, vesicular basalt, basaltic glass, scoria), andesite, dacite and rhyolite (as vitrophyre, obsidian).
- A set of rock-forming minerals that include quartz, sodium feldspar, potassium feldspar, muscovite mica, jasper, calcite, olivine, iron, amphibole, pyroxene, biotite mica and calcium feldspar

Please note: When talking about the dark-colored rock-forming minerals, I use a dark blue color for the text. When talking about the light-colored rock-forming minerals, I use a lighter blue color for the text. These colors are part of a color code for rocks and minerals that I have created that exists across the different books that I have written. Volcanic rocks in this text will appear in a dark orange. Metamorphic rocks are red, plutonic rocks are blue, and sedimentary rocks are brown.

When **Word Challenges** first appear in the text, they will appear ***bold and italicized.*** **Be sure to write out the meanings of each of these words in your lab book.**

Lesson One
The Framework for Interpreting The Hawaiian Islands

Word Challenges: *assumption (assume), catastrophism (catastrophic), channelized erosion, chronology, framework, history, naturalistic (naturalism), philosophy, secular, sheet erosion, uniformitarianism, worldview*

The word *geology* means, *the study of the earth.* It includes both the science *(Earth science)* and the **history** *(Earth history)* of our earth. One of the best places to study geology is in our National Parks.

The National Park system – what a great idea! It has been the model for all nations since the inception of the first National Park – Yellowstone National Park in 1872. And the Hawaiian Islands contain several volcanic sites that have been preserved as either National Parks or National Monuments. The American National Park system is the envy of the entire world, as people have been dedicated to preserving the wonders of nature for future generations to visit and study. And this was the intent when men first petitioned Congress to protect the wonders of the Yellowstone region. It remains the primary goal of the National Park system today. I am so thankful that American people over 100 years ago had the foresight to commit themselves to this ideal. But how do we study these parks? Most people are not equipped to study the Parks from any perspective other than that which takes God and His word out of the mix. The goal of studying God's creation is to be reminded of who He is. But how can we do that if we are constantly bombarded with thoughts and doubts that are contrary to God's recorded history?

What is a Framework?
To study the Islands, we need to understand what framework we are going to use. What is a **framework**? A framework is an interpretation of evidence. Every human being uses a framework when thinking about life, science, history and philosophy. For instance, your view of Cambodia today might be influenced by whether or not you had family living there in the 1970's who might have suffered terrible persecution. This could impact how you think or relate to Cambodia today – this is your framework. But did you know that *every scientist* also uses his/her framework to formulate ideas and opinions about physical evidence? And scientists are often unaware of just how their framework affects their conclusions. A good scientist will be aware of his framework and will be careful to keep that framework from influencing the results of research.

Sadly, the interpretation or framework of America's landforms that most scientists use today has been taken over by a **naturalistic worldview**. That means that all the natural wonders enshrined in our national parks have been explained from a **secular** perspective so that the Biblical history of the earth is obscured. Many people are stumbled by this viewpoint such that the book of Genesis is becoming a just-so story, not rooted in reality. Most people don't give it a second thought. However, the implications for the Christian are enormous. Everything that has been taught about Jesus Christ is rooted in the book of Genesis. If Genesis is false, then the foundation for Jesus as our Messiah and savior is at best suspect, and at worst, a lie.

Understanding Uniformitarianism
Early in the development of modern geology, men adopted a naturalistic framework for interpreting the landforms of our earth called **uniformitarianism.** This initially meant that the geological forces that are shaping our earth today have been going on almost indefinitely. A phrase used by biologist Dr. Gary Parker states this view in a way that is easy to remember – *"Small and slow and long ago."* Although the term has been slightly modified today to include localized geological catastrophes, like volcanic eruptions, the concept of an ancient earth, perhaps as old as 4.6 billion years, has remained. This means that when a secular geologist looks at the Hawaiian Islands, for example, because he sees slow and gradual erosion in operation today, he **assumes** that this geological force has been going on for hundreds of millions of years. Those who formulated modern geology in the early 1800s made a huge historical and consequently a huge geological mistake. They rejected the global flood of Genesis. If indeed a global flood had occurred as Genesis records, this concept would have totally changed the course of modern geology. The global flood of Genesis would have produced the effects highlighted in another one of Dr. Parker's sayings, *"Big and fast, and in the recent past."* These two ideas are diametrically opposed to each other.

Each one of the Hawaiian Islands are most often interpreted through the secular, uniformitarian framework to tell the public a naturalistic story with a naturalistic beginning.

Catastrophism – a Biblical Framework
You will find it makes a tremendous difference when a person takes Genesis as recorded history and interprets the formation of the Hawaiian Islands through

this framework. The Bible seems to come alive and the reader is struck with a fresh sense of the historicity and reality of the Biblical record.

In contrast to the uniformitarian framework involving small and slow geological processes over millions of years, is the Biblical view or the *catastrophic* framework of Earth history. The first 11 chapters of the Bible give a concise and historical framework through which the landforms of the earth can be interpreted. It is best summed up in the following diagram.

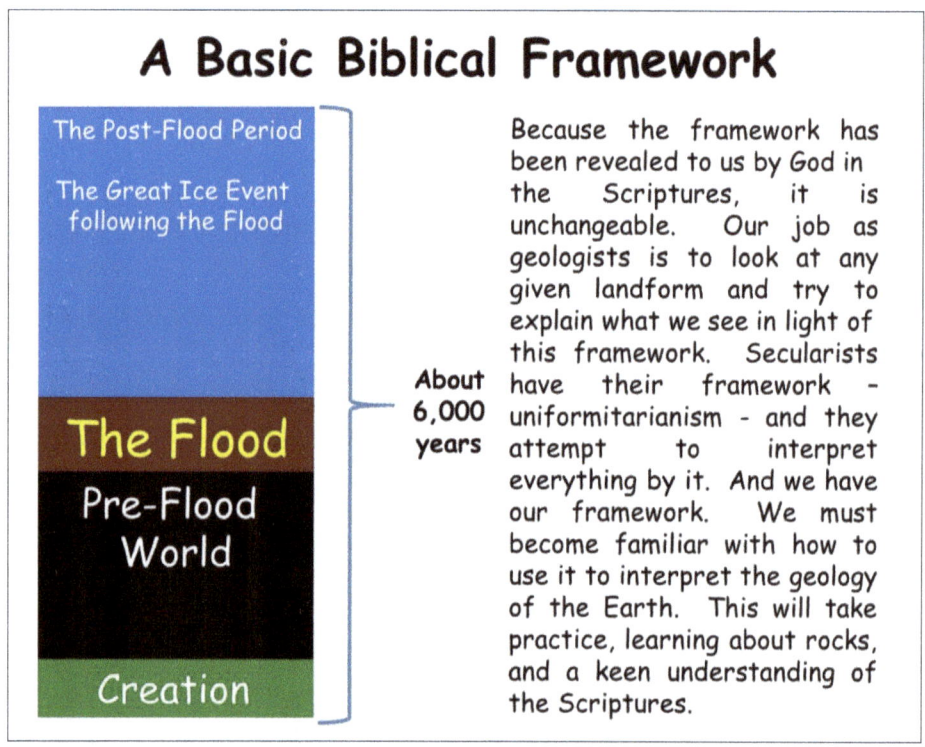

In this framework, we don't start with evidence, because a person will always be influenced in the interpretation of that evidence by his philosophical framework. Instead we start with a recorded history of the earth. The Biblical, historical framework has been tested historically, geographically and archaeologically time and time again, and has stood the scrutiny of thousands of people through the years. This reliable framework is what we will use to interpret the evidence. It helps us to fit what we observe into a more historically complete picture. Modern geology tells their own historical story in interpreting the geology of our National Parks and then weaves evidence into that story. And it is an always changing story, from year to year, as popular explanations fall out of favor, to be replaced with another idea. There is no recorded history to rely on. This will almost always be contrary to the Biblical story. And it is this secular approach that has caused so much confusion and consternation for those who believe the Biblical story.

The Bible also reveals a historical ***chronology*** of the Flood in Genesis chapters 7-8. Some may offer a little different analysis of the number of days involved in the Flood period, so the following two charts are simply my attempt to promote a general geological structure of the period known as The Flood. The first chart is more comprehensive while the second is a summary. Please take time to study the scripture references on the first chart.

The Genesis Flood – Order of Events and the Geological Implications
(not to scale) Genesis 6:9-8:16, beginning about 4,500 years ago

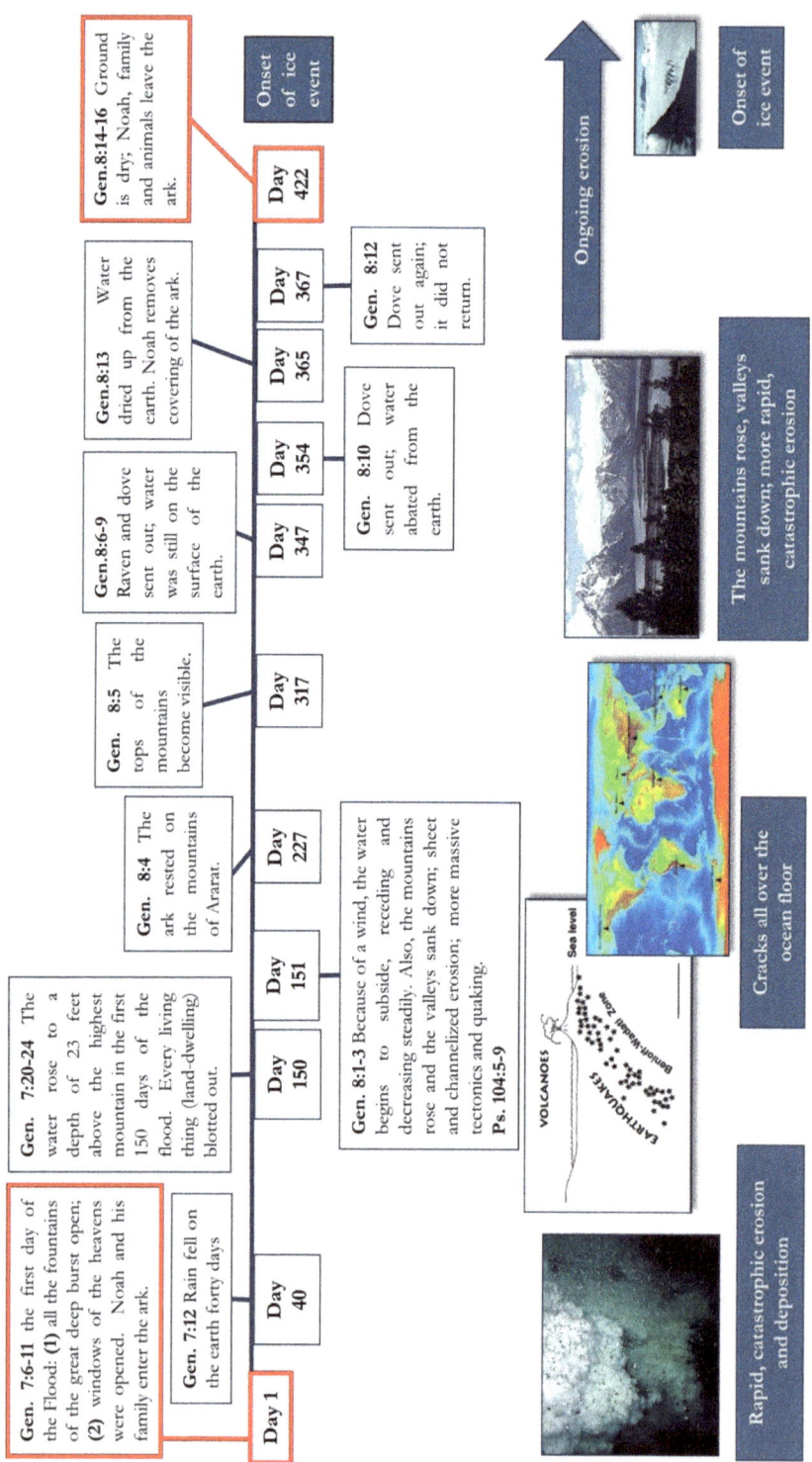

We will use The Basic Biblical Framework and this chronology as we look at and seek to interpret the geology of Hawai'i.

To sum up the preceding chart, here are the primary points.

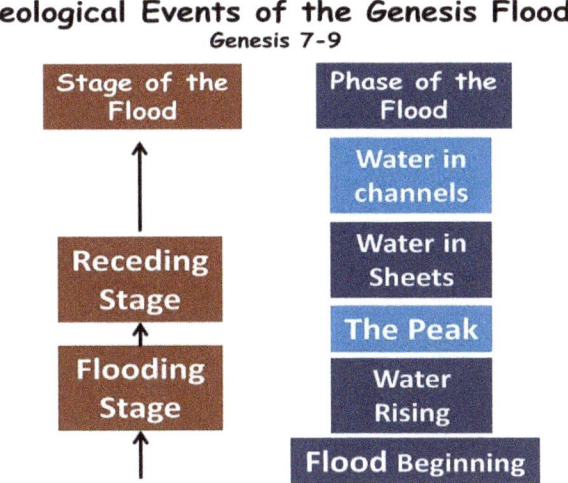

Sheet and Channelized Erosion

When we study geology, not just of the Hawaiian Islands, but anywhere, we need to realize the impact of the receding stage of the Flood. The recession stage of the flood beginning at day 151 involved two phases of rapid and massive erosion – **sheet and channelized erosion.** Sheet erosion is when water comes off of the rising mountains in sheets, and consequently planes surfaces, producing sheared, flat surfaces; channelized erosion happens when this flow decreases, producing canyons and ravines. Read Psalm 104:5-9. We will see some of these effects as we study the geology of Hawai'i.

Evidence is only what we can observe now, in the present. To put a template over the evidence so that it portrays a history must be done with a worldview, a framework. There are only two frameworks that could interpret the evidence. Uniformitarianism – *"Small and slow and long ago,"* or, **catastrophism** – *"Big and fast and in the recent past."* There are really no other frameworks that could put the geological pieces together.

Thought Questions

1. Define the word, *framework*.

2. Define the word, *uniformitarianism*.

3. Define the word, *catastrophism*.

4. Define the word, *secular*

5. Dr. Gary Parker uses two sayings to summarize the two opposing frameworks used in interpreting the geology of the earth. What are they? Can you describe in your own words, what they mean?

6. Briefly list the four main parts to the Basic Biblical Framework.

Activity: Memorize the Basic Biblical Framework and then draw it from memory. It is vitally important to know this framework if we are to understand the geology of the Hawaiian Islands.

Lesson Two
Radiometric Dating and the Hawaiian Islands

Word Challenges: *archipelago, atoll, atom, atomic mass,* basalt, dacite, *element, geologic column, half-life, isotope, lava, periodic table, radioactivity, seamount, volcano*

What are the Hawaiian Islands?

The Hawaiian Islands are a chain of **volcanoes** that stretch about 1,500 miles in a northwest direction, from the Island of Hawai'i to Kure **Atoll**. Each one of these islands is a volcano or a complex of volcanoes of different types. The state of Hawai'i includes this entire chain except for Midway Island, which is an unincorporated territory. The technical name for a group of islands is **archipelago.** Sometimes it is called an island group or island chain. The word, archipelago, can refer to any island group; it can also refer to a sea containing a small number of scattered islands.

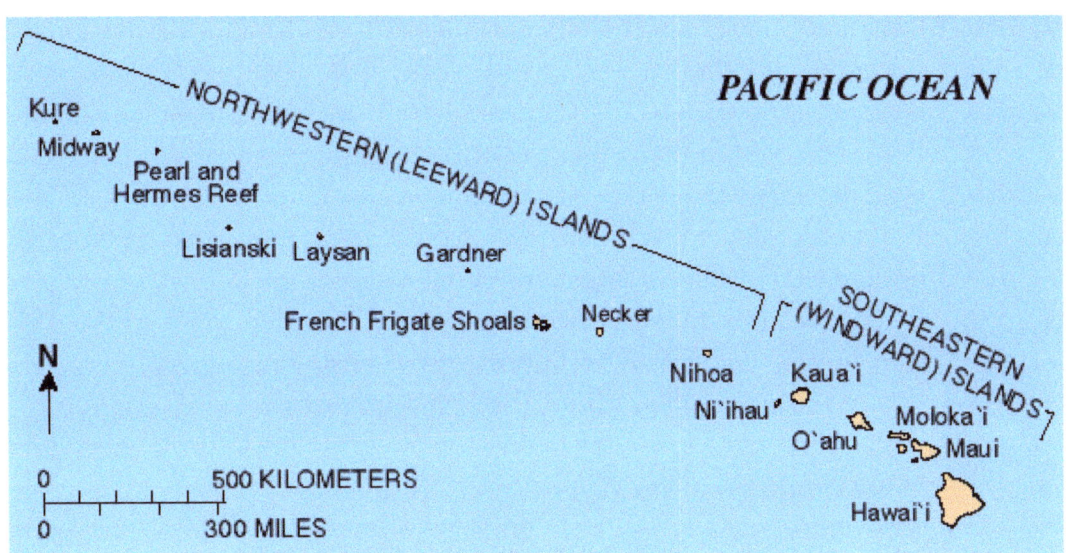

The Hawaiian Islands are a chain of volcanoes.

Geologists *estimate* the age of these islands from *over 50* million years old for the northwestern most island, Kure Atoll, to 400,000 years old for the southeastern most island, Hawai'i. The island of Hawai'i still has active volcanism, as does the submerged but growing underwater volcano, to the southeast of Hawai'i, Lo'ihi. Lo'ihi is called a **seamount**, a type of underwater volcano. All of the Hawaiian volcanoes have produced or are producing, **lava**. As it is important in any study of the Hawaiian Islands to understand lava, let's look at it.

Lava is a word used to describe molten rock that comes out of a volcano or fissure. It's also used to describe the resultant rock that is formed by this. Geologists use the potassium-argon radiometric dating method to *estimate* the geologic age of the lavas on each of the islands. Geologists use terms such as *estimate* and *approximately*, because the radiometric "clocks" are not totally accurate.

We will study the different lavas of the Islands later in the book. At this point, it is enough to know that much of the dating of the Islands is based on the supposed radiometric age of these lavas. We will also take a look later at the use of radioactive **elements** to estimate age. But first, let's look at radiometric dating.

A Short History of Radiometric Dating
Part of the Age of Enlightenment (also called The Age of Reason, late 1700's-early 1800's) involved rejection of the Biblical history that included the Genesis account of Creation and the Global Flood. This was not done based on any scientific discovery, but purely through a change in philosophical thinking. "Enlightened" people would no longer depend on religious stories or myths, like the Bible. And because the Bible taught a recent creation and flood, these "enlightened" thinkers developed ideas that would go in the opposite direction, involving hundreds of millions of years of slow geologic development and evolution. Today, this thinking has become science doctrine. But in the 1800s the ages chosen for the geological evolution of the earth were guesses and they were constantly changing as ideas about the beginning of the earth changed.

Then in 1896 **radioactivity** was discovered. It seemed to show that certain elements decayed over time – some very quickly and some very slowly, seemingly on the order of hundreds of millions of years. Geologists began to consider this radioactive process as a possible way to objectively and scientifically date the age of the earth. This process became sort of a clock and has become known as radiometric dating. It seemed like this would be the final nail in the coffin of Genesis and its teaching of a recent creation and flood.

What is radiometric dating?
Radiometric dating is a system that uses the process and known present rates of radioactive decay to project the age of a rock provided certain assumptions are accepted.

What is radioactivity?

Radioactivity is the tendency for certain elements, called *isotopes*, to lose energy. The loss of energy from an isotope is measured as a decay per second, called the Becquerel, after its discoverer Henri Becquerel in 1896. Elements that decay are radioactive.

What is an isotope?

The word isotope is from the Greek word that means, *same place*. An isotope occupies the *same place* on the **periodic table** as its counterpart. For example, the regular carbon **atom** (C-12, or carbon-12) is found in the 14th Group. It is considered to be stable. Carbon-14, one of its isotopes, would occupy the same place, but it has something different about it. Let's look at what that means.

The periodic table was developed by Dmitri Mendeleev in 1869; only 63 elements were known at the time

Atomic configuration

Elements are specially designed atoms, all having a specific number of atomic particles called protons (+, a positive charge), neutrons (0, neutral, which form the nucleus of an atom) and electrons (-, a negative charge). The elements are all uniquely and specially configured to give each element its characteristics. The elements make up the periodic table. Changing the number of these particles changes the element entirely. In the periodic table, each element is configured as a square with letters and numbers. The number in the upper right-hand corner of the square is the atomic number. It tells us how many protons and electrons an atom has. In the example of carbon-12, there are six protons and six electrons. The number of electrons always equals the number of protons. The letter(s) in the center of each square is the

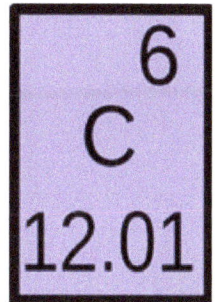

chemical symbol for the element – in this example, carbon. The number on the bottom represents the ***atomic mass***, which equals the combined number of protons and neutrons. So, how many neutrons does a normal carbon atom have? The answer is six. In carbon-14, however, the atomic mass is 14, which means that it has 2 more neutrons than normal carbon and therefore is unstable. It decays. It loses energy. It is radioactive.

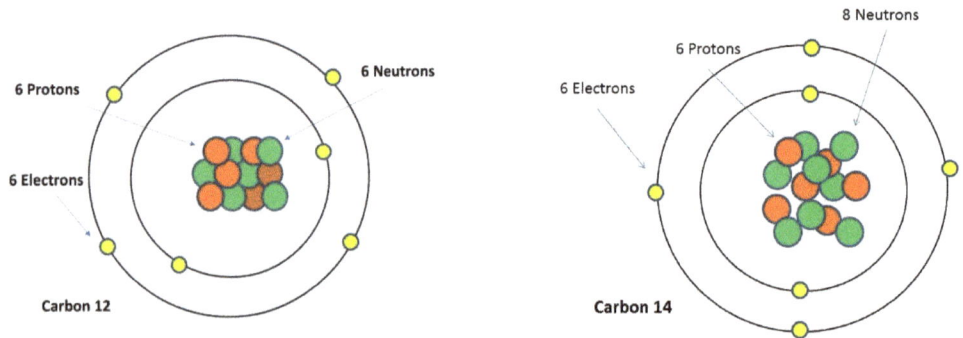

Configuration of carbon-12 and carbon-14

Some radioactivity is extremely dangerous. Some is not. The important thing to remember is that radioactivity can be measured. No one has ever been able to definitively say why radioactive elements exist.

In light of the fact that we have radioactive elements in our universe, here is an important question for those who believe in the Genesis account of creation and the flood. Why, if after the sixth day of creation, God pronounced everything, *very good*, do we have abnormal and dangerous radioactive atoms in the mix of the original atoms that God created? No one has ever been able to scientifically show why we have abnormal atoms. Scientists have been able to configure all of the naturally occurring atoms and their structures. And we know just how much of the elements our earth, and life on it, has. And each normal element is necessary in just the right amounts – even arsenic! It might very well be that radioactivity was a vital part of the inner workings of the earth but nonexistent on the surface of the earth at creation. At the point the fountains of the great deep burst open, in Genesis chapter 7, radioactivity might have been released in great amounts producing the false readings that are so abundant today. The origin of radioactive elements is not well understood.

What happens to a radioactive element?

Because an abnormal element has an abnormal number of atomic particles, it tends to "shed" those atomic particles with the net effect of changing into a stable element. Because these atomic particles can be identified and observed, a statistical prediction can be made. For example, if carbon-14 loses particles at so many Becquerels or decays per second, then theoretically it should turn into another stable element, given certain assumptions and enough time.

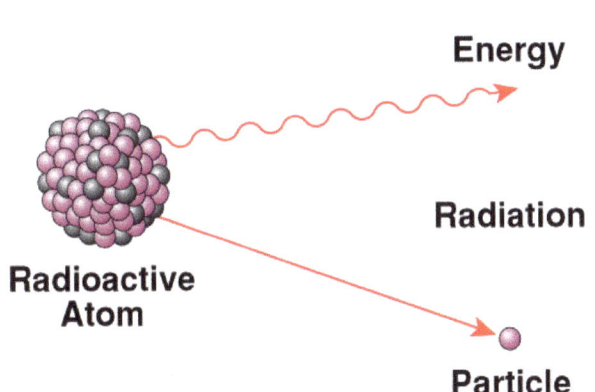

We will discuss these assumptions later. Notice in the following example, in the process of radioactivity, an atomic particle is released. There are several different kinds of these particles depending on the type of radioactivity. The releasing of this particle is energy released in the form of a Becquerel, or decay per second, and these can be measured as they are observed.

Radioactive decay

If radioactive elements (atoms) lose particles in the decay process, what happens to the original radioactive element? Obviously, there is going to be a change that takes place. The radioactive element theoretically becomes something else and loses its own identity, so to speak. I say theoretically, because the decay process can require an immense amount of time to complete and would not be able to be observed throughout the entire process. The unstable element that is decaying is called the *parent* and the stable element into which it theoretically eventually changes is called the *daughter*. Many of the radioactive elements used in radiometric dating require huge amounts of time to decay and no one has recorded or witnessed the entire decay process. For example, the time it takes for ½ of carbon-14 to decay, measured at present measurable rates is a little over 5,000 years. Then in another 5,000 years ½ of what is left decays and so on until it is all gone. This is what we refer to when we say, **half-life**. Has anyone ever observed the complete process to know whether it has actually completed this process or not? The emphatic answer is, "No." The time involved for this to take place is far beyond human witness. Since science is the study of what can be observed then this process is outside the realm of science.

In this case, it would take over 100,000 years! The following is an illustration of the idea of half-life. Theoretically all that is left at the end of the decay process is the stable daughter element.

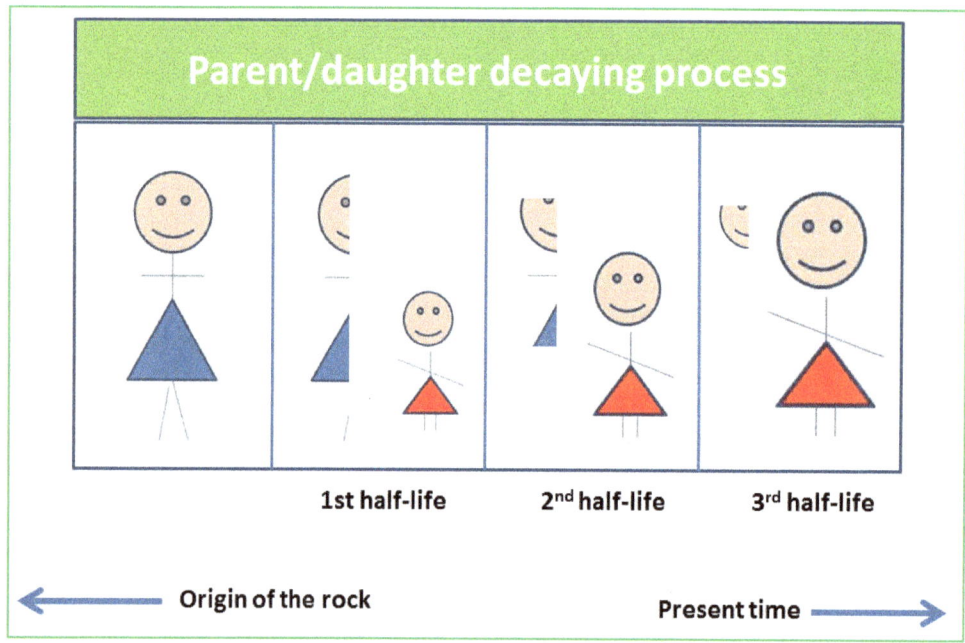

The present Becquerel determines the half-life of the radioactive element. The time is then projected forward or backward to arrive at a time when the process theoretically started or when it would theoretically end. In our carbon-14 atom above, the stable element, into which C-14 would decay, is nitrogen-14. This is a chemical prediction based on what we know of present radioactive decay. It is a statistical arrangement of the parent particles that points the way to the daughter element. Remember that the process, if totally accurate, would take thousands of years to complete. No one has been around that long.

Uranium decay

Radioactive Uranium-238 (U-238) is a very heavy element. But notice its half-life from the following chart. Has anyone ever observed this? The answer is obviously, "No". Then how do we know that it operates this way? We can *guess* based on the present chemistry of radioactive decay. It is a matter of statistics. If U-238 were to continue to lose particles like it seems to do in the present, then over such and such a length of time it will go through a chain of decays and changes until a stable element is arrived at – in this case, Lead-206.

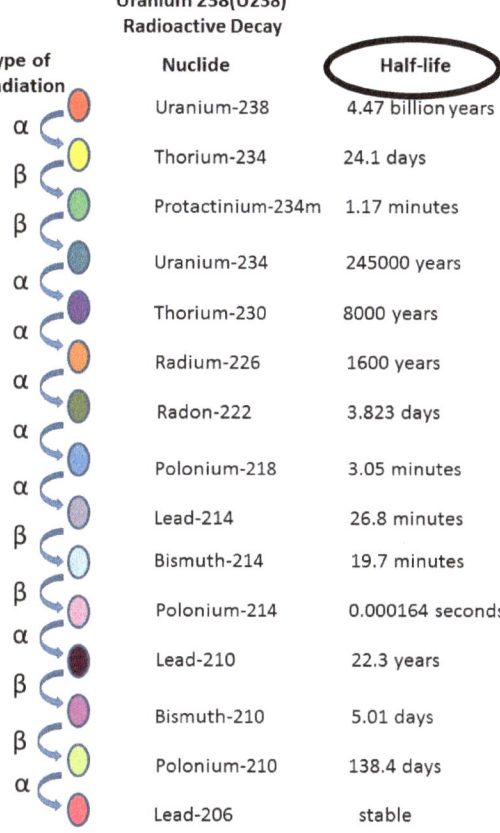

Potassium/Argon Dating Method

In addition to the uranium/lead, scientists also use the popular potassium/argon method (K-40/Ar-40). It works exactly the same way. The uranium/lead (U-238/Lead-206) method is a commonly used method in calculating the radioactive decay/age of a rock. But if no one was around to actually test or record the entire process from beginning to end, then how do we know this process works? We don't! Geologists make *assumptions* based on the present observation of radioactivity. Does that process sound familiar? It should. It is an application of the philosophy known as uniformitarianism – the present is the key to the past. So, by assuming that the present observable process has always been true in the past, geologists can use the process of radioactive decay to date rocks. How do they do this?

Only rocks that are thought to have been molten at one time are considered to be valid specimens for radiometric dating. Volcanic rocks are mostly used. But at times, attempts have been made to date plutonic and metamorphic rocks too. Geologists also consider these rocks to have been molten at one time. We will discuss this later when we discuss rock types.

The Geologic Column or Time Scale or Time Table

Study the following chart, which is known as the *geological column, time scale or time table*.

EON	ERA	PERIOD	MILLIONS OF YEARS AGO
Phanerozoic	Cenozoic	Quaternary	--- 1.6 ---
		Tertiary	--- 66 ---
	Mesozoic	Cretaceous	--- 138 ---
		Jurassic	--- 205 ---
		Triassic	--- 240 ---
	Paleozoic	Permian	--- 290 ---
		Pennsylvanian	--- 330 ---
		Mississippian	--- 360 ---
		Devonian	--- 410 ---
		Silurian	--- 435 ---
		Ordovician	--- 500 ---
		Cambrian	--- 540 ---
Proterozoic	Late Proterozoic / Middle Proterozoic / Early Proterozoic	Ediacaran 635-543 MYA	--- 2500 ---
Archean	Late Archean / Middle Archean / Early Archean		--- 3800? ---
Pre-Archean			

Extinction Event 65 Million Years Ago (End of Cretaceous) – dinosaurs and mammals

Extinction Event (End of Permian)

Cambrian - 540 Million Years – marine creatures

It is presented as a scientific fact in every book about geology. It usually shows small marine creatures on the bottom, progressing to fish, amphibians, reptiles, and finally large land animals, and man. But in reality, it is a hypothetical idea. The Column in its *entirety* has not been found anywhere on the earth. Some of the rock layers do occur in an apparent order, but there is a Flood explanation for this apparent order that modern geologists reject. It is a general order (with many exceptions) produced by the sorting action of the Genesis Flood. If the fountains of the great deep burst open first, as Genesis states, then one would expect that marine (sea) creatures would have been the first victims of the Flood. As the earth became more inundated with water, other creatures would have followed as the Flood swallowed them up. Man, being the most mobile, would have most likely been buried last, if at all. Land creatures swimming in order to survive would have ultimately drowned, and fish, sharks, or other scavengers would have eaten their remains.

What about fossils? How do we date them?

At this point, we need to talk about fossils and how they are dated. *Fossils cannot be directly dated radiometrically because of contamination.* They are a mix of organic and inorganic materials. Fossils are dated by another means - evolutionary belief – using uniformitarianism. That is, creatures change in the present. Therefore, creatures have changed in the past and given enough time, have changed and will change radically into other types of creatures – *evolution*.

As the idea of change in creatures expanded to an evolutionary view where one type of creature was thought to have changed into an entirely different creature, so the amount of time for nature to accomplish this was also necessarily expanded from several thousand years to 550 million years by the end of the 19th Century. The basic geologic column used by scientists today was set by the mid- to late-19th Century. In other words, the concept of time plus evolution was a philosophical shift from a Divine Creation taught in the Book of Genesis, to a totally naturalistic and atheistic one taught through uniformitarianism, not science. Today many people think that the geologic column and the millions of years it involves is a scientific fact supported by radiometric dating. Remember, however, that radioactivity was not even discovered until after the geologic column had been in place for many years. Most people are unaware of this. It was not science that formulated the geologic column, but philosophy. It was a naturalistic attempt to explain the apparent order of fossils. That is, marine fossils on the bottom of layers, dinosaurs closer to the top. It is an idea, and an accepted idea, but not proven science. Secular scientists, therefore, determine how old fossils are by where in the geologic record that they are found.

How is a radiometric date arrived at?

Let's start with a volcanic rock. And let's say it is chemically analyzed to contain 80 parts of uranium and 20 parts of lead. Is the rock young or old? Remember what we discovered about uranium decay? Well, based on what geologists have assumed about the decay of uranium to lead, and since more of the uranium is present than lead, then the rock must be young. If the ratio was reversed, then the opposite would be true – *given the following* **assumptions**:

1. We *assume* that the initial state of the rock started with a certain amount of uranium and no lead. In other words, the initial state of the rock is *assumed*.
2. We *assume* that there was no lead present at the start of the process.
3. We *assume* that no uranium came from some other source.
4. We *assume* that all of the lead that is present came from the decay of uranium and that it did not come from some other source.

5. We **assume** that the decay rate or process has not been interfered with from some other means or sources.

So long as we go with these *assumptions*, we can statistically figure an age for the rock.

But what happens when different ages are obtained, as in the case of Mt. St. Helens lava, known to be ten years old, but which dated in the hundreds of thousands of years? Scientists assume that the radioactive decay of a sample begins when the lava cools. Again, geologists simply *assume*:

1. You or the laboratory contaminated the samples.
2. You or the laboratory made a mistake in calculations.

Mt. St. Helens Sample	Age in Millions of Years
Whole dacite lava rock	0.35 (350,000 years old)
Feldspar from the dacite lava rock	0.34 (340,000 years old)
Amphibole from the dacite lava rock	0.90 (900,000 years old)
Pyroxene from the dacite lava rock	1.7 (1 million, 700,000 years old)
Pyroxene sample #2 from dacite lava rock	2.8 (2 million, 800,000 years old)

Potassium-argon 'ages' for whole rock and mineral concentrate samples from the lava dome at Mt. St. Helens that was in reality about 10 years old.

1981 - The new *dacite* lava dome in Mt. St. Helens' blown-out cone, from which the samples were taken

Other volcanic rocks from fairly recent recorded eruptions that were witnessed, also show radiometric discrepancies:

Date Eruption Witnessed	Rock/Mineral Sample Dated	Radiometric Date
Hualalai – 1800-1801	Basalt	1 million 600,000 years old to 410,000 years old
Mt. Etna – 122 BC	Basalt	250,000 years old to 80,000 years old
Mt. Etna – 1792	Basalt	350,000 years old
Mt. Lassen – 1915	Plagioclase Feldspar	110,000 years old
Sunset Crater – 1064-1065	Basalt	270,000 years old to 250,000 years old

Other volcanic rocks from eruptions on Hawai'i illustrate the same pattern of discrepancies in the radiometric dates:

Rock Sample Obtained From:	Known Age from Historical or Archaeological Data of Eruption	Rock's Age from Radiometric Dating	Dating Method Used
Hualalai Volcano	200 years	140-670 million years (range of dates)	Escaping Helium
Hualalai Volcano	200 years	160 million years to 2.96 billion years (range of dates)	Potassium-Argon
Kilauea	200 years	-0- years at 400 foot depth	Potassium-Argon
Kilauea	200 years	10-14 million years at 10,260 foot depth	Potassium-Argon
Kilauea	200 years	13-29 million years at 14,040 foot depth	Potassium-Argon

But what if the sample from the same geological area produces different ages using *different dating methods*, as in the case of a *basalt* sample from the Grand Canyon where different ages for the same rock were obtained using *different* methods? Geologist Dr. Steven Austin sampled basalt from the base of the Grand Canyon strata and from lava that spilled over the edge of the canyon. By evolutionary reckoning, the latter should be a billion years younger than the basalt from the bottom. Standard laboratories analyzed the isotopes. The rubidium-strontium isochron technique suggested that the recent lava flow was 270 million years older than the basalt beneath the Grand Canyon – an impossibility.

Dating Method Used	Derived Radiometric Age
6 potassium-argon model ages	10,000 years old to 117 million years old
5 rubidium-strontium method ages	1,270 years old to 1.3 billion years old
Rubidium-strontium isochron method	1.34 billion years old
Lead-lead isochron method	2.6 billion years old

Radiometrically derived ages, using different methods, for basaltic rocks most geologists accept as only thousands of years old, from the Uinkaret Plateau of the Grand Canyon

The umpire in cases like these is the Geological Time Scale (Geologic Time Table). Since dates to 550 million years old were worked out in the 1800s, and have already been agreed upon, if the date for the rock appears to be too old or too young, then the dates are either thrown out or selected. The date, which is the closest to what the geologist thinks it is, becomes the accepted date. *And that's radiometric dating.*

How does all this connect to the Hawaiian Islands?
One of the characteristics of the magma (lava) that erupts on these islands is that it is very rich in radiogenic argon (an isotope of argon). Samples from the Islands regularly give ages that are far too old for these recent flows, most likely because of the abundance of the argon. There is too much radiogenic argon gas. There should be a lot less for the supposed radioactive process.

Radiometric assumptions are not reliable
What might have caused the rate of decay to change in the past? The Genesis Flood would have added a tremendous amount of heat to rocks. It would have also added a tremendous amount of hot water to the rocks. And in fact, it has been shown that radioactive decay was rapidly sped up in the past in several substances:

1. Carbon-14 was discovered in diamonds thought to be at least several billion years old. According to present measurements of the decay of C-14, there should be absolutely no detectable C-14 after approximately 100,000 years.
2. Carbon-14 was discovered in petrified wood thought to be at a few hundred million years old.
3. Recent observed lava flows in New Zealand and in Hawai'i dated excessively old according to modern radiometric dating methods.

So let's think about this: if radiometric dates need to be checked by another source, the Geologic Time Table, for accuracy and reliability, then, are they reliable methods?

These examples show that the radioactive decay has been affected by some process. Could it be possible that the Flood impacted our earth in such a way that radioactive processes were radically affected? Flood geologists suspect this could be the case. This would mean that thousands, if not hundreds of millions of years of radioactive decay could have occured in a matter of a few months. Even though radioactivity is an observable fact, the measurement and predictions of this process have been affected by something else and should not be used as a basis for age/dating. It would be like hanging on to a clock or watch that ticks, but is always gaining or losing time. Why depend on it? It's a clock, but is it reliable?

Mt. Ngaruohe, New Zealand: samples from here all gave conflicting radiometric ages

Why do geologists continue to use an unreliable method?

Geologists continue to use the radiometric dating methods despite the obvious flaws. Why? Although there is rarely an open public admission of these errors, the radiometric dating methods provide the only "scientific" alternative to the young-earth conclusions of the Bible. At the center of the radiometric dating rational are the influence of The Enlightenment and consequently the extremely narrow view of uniformitarianism. Anything but the Bible!

Has this ever been observed? How could it? The creation itself is only about 6,000 years old! In addition no one has ever been around long enough to have recorded it. Scientists simply assume that given the observable decays per second and enough time, such and such would happen. It must be remembered here that *radioactivity* (the tendency for isotopes to lose energy and consequently atomic particles) is a fact of science. *Radiometric dating* is not a scientific fact but an application of the scientific fact of radioactivity. Radiometric dating is based on the philosophy of uniformitarianism. As it applies to radioactivity, uniformitarianism would state that what we observe in the present has always been going on in the past at the same rate and in the same way we observe now. And therefore, given enough time, an isotope would transform into a stable element. This sounds scientific and reasonable. However, if we introduce a historical global flood into the picture, then uniformitarianism is a false premise. Also, a global flood could have changed or interrupted normal decay rates and even introduced contaminates into otherwise natural processes.

Recent lava flow in Hawaiʻi: samples all gave conflicting radiometric ages.

Thought Questions

1. What is an atom? What is radioactivity?

2. What is radiometric dating? How is it supposed to work? What must be accepted in order for radiometric dating to work?

3. What kinds of rocks are normally used in the radiometric dating process?

4. Name at least four assumptions that are used to arrive at a radiometric age.

Activity: Look up four different radiometric dating methods. List the atomic configuration for each of the elements involved, both normal and isotope. Name the parent element and the daughter element in each dating method chosen.

Lesson Three
The Origin of the Hawaiian Islands

Word Challenges: *caustic, corruption, deterioration, lahars, pH scale, pyroclastic, science*

The Age of Hawai'i

The age of the most northwest island in the Hawaiian chain of islands, Kure Atoll, is given as roughly 28 million years old, and the main part of Hawai'i is aged at roughly 400,000 – 5 million years old, according to secular geologists. How did they come up with this figure? By applying the philosophy of uniformitarianism to present observable radioactive decay rates, geologists have been able to estimate the ages of all kinds of places. The way the ages are quoted allows for very little disagreement. But since the radiometric dates appear to be unreliable, is there any other scientific way to age or date the remote geological past? No, there are none. Secular geological history is outside the scope of observable historical records. Then how can this area of geology be *science*? I was always taught that science involved that which was observable, testable, and repeatable. It is not possible to observe anything or repeat anything to determine that the Hawaiian Islands are 28 million years old. And given the unreliability of radiometric dating methods, neither is it testable.

Now you may not think this is a huge issue. It may not even matter to you. But let's think about this.

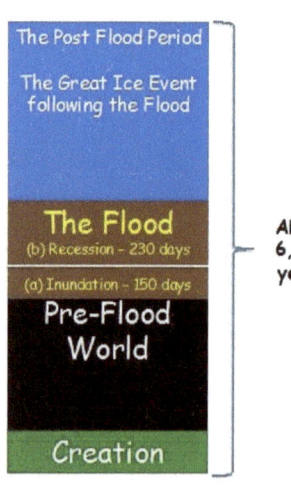

A straightforward reading of the Bible clearly teaches that the earth is about 6,000 years old. It also clearly teaches that a flood of global proportions totally rearranged the surface of the earth. This is not a scientific observation. It is, however, a matter of historical record. If it is true, then there is absolutely no one way for secular geologists to be right in their declaration of 28 million years old for a Hawaiian island! If secular geology is right, then there is no way that your Bible can be true – no ifs, ands or buts. The issue is that clear. So, we have to look at our Basic Biblical Framework, discussed in Lesson One to resolve this dilemma. Since we know that the issue cannot be decided by science, then we must look to history and our Biblical historical record.

The Creation

The last verse of Genesis 1 states, *"God saw all that He had made, and behold, it was very good. And there was evening and there was morning, the sixth day." (Genesis 1:31)* This took place before the **corruption** and **deterioration** of the earth was brought in by man's sin. What was Earth like before sin and corruption? We don't fully know, but it must have been a wonderful place without sin, without decay, without corruption, without death and without disease and sickness. Hawai'i is a chain of volcanoes. So, the question we must ask here is this. Were volcanic eruptions part of the original creation? Did they help form the original creation? And how old are the Hawaiian Islands? In order to answer these questions, we need to look at the nature of volcanoes.

What comes out of volcanoes?

According to the United States Geological Survey, here are the main things we need to be worried about in a volcanic eruption:

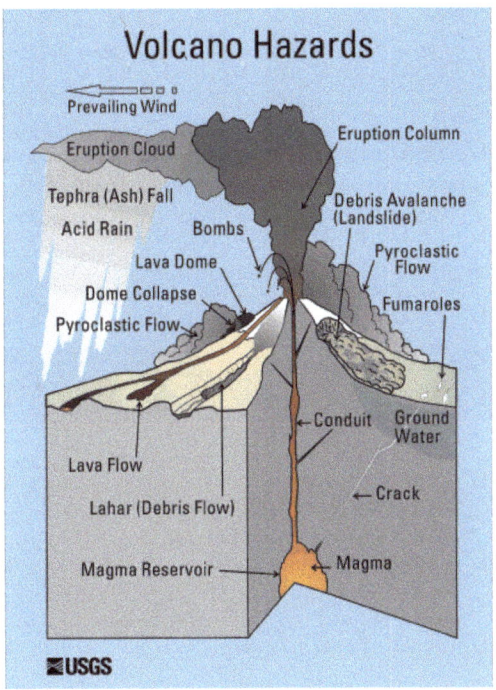

a) **Acid rain** – This is acidic water from the steam that is part of a volcanic eruption. The acid is primarily sulfuric acid, a very **caustic** and destructive acid. Take a look at the *pH scale* for measuring acids and bases. Volcanic aerosols contain a good deal of caustic stuff. Acid rain has a pH of 5.0 or less.

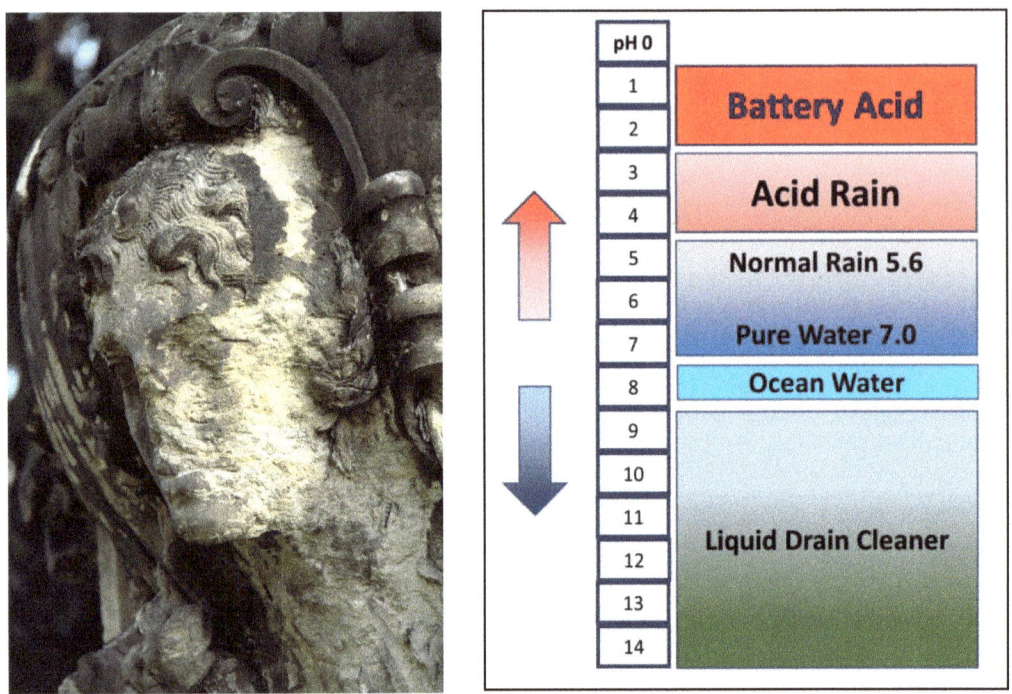

(Left) This statue shows the effects of acid rain. (Right) pH scale

b) **Debris avalanches** or *lahars* (volcanic rock and debris landslides) – These alone have been responsible for destroying towns and villages.

(Left) The aftermath of a lahar from the 1982
eruption of Galunggung, Indonesia
(Right) The lahar from the 1985 eruption of Nevado del Ruiz
that wiped out the town of Armero in Colombia

c) **Ash fall** – This is not like the ash that comes from burning wood or paper.  Ash fall is a mixture of bits and pieces of old volcanic rock and glass. Ash destroys machinery and causes lung disease in animals that breathe the ash, and eat the plants on which the ash falls. The ash clouds can travel in the atmosphere for days or weeks, circling the globe many times. The bones in this picture show lung failure due to breathing in volcanic ash. The fossil is of a *Teleoceras* – rather like a rhinoceros: these fossils are abundant in Nebraska.

d) ***Pyroclastic*** **flows** – These are very hot gas clouds filled with glass, steam and bits of older volcanic rock that hug the ground, moving along at speeds of up to several hundreds of miles an hour and stripping everything in their paths. The trees in the following photo were all broken like tooth picks and laid down like pick-up sticks as a result of the pyroclastic flow of Mt. St. Helens.

(Left) Felled trees at Mt. St. Helens from a pyroclastic flow (Right) An ash cloud pyroclastic flow, Philippines 1984

e) **Lava flows** – These destroy everything in their paths, including towns roads, and vegetation.

Lava flows on the Big Island of Hawai'i

f) **Noxious gasses** – Carbon dioxide and sulfur dioxide are deadly gasses, and they are part of the gases that are emitted in an eruption. They not only choke living animals, but when mixed with water, they produce acids. The diagram below shows the aerosols and gases that can be emitted in the atmosphere and consequent damage, including acid rain and ozone destruction.

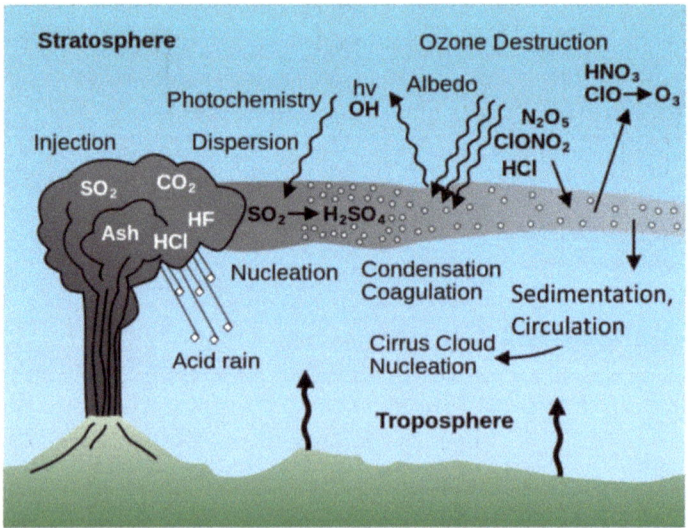

Not only are there immediate hazards from volcanic eruptions, but also terrible scars are left on the earth in the aftermath. Study the following pictures and notice some of the vestiges of volcanic eruptions.

(Top) Erosion; lahar damage (Bottom) ash covering the field 180 miles from eruption; boulder moved by lahar

Of course, God watches over His earth and continues to care for the people and the planet. Volcanic rock breaks down to form and nourish soil. And in a sense the earth is partially renewed. This can be seen in the aftermath of the Mt. St. Helens eruption. What was once a barren landscape has regained some of the beauty that reflects the Creator.

Mt. St. Helens, pre-eruption   **Mt. St. Helens, post-eruption**   **Mt. St. Helens, after 20 years**

Here is a list of some of the potential destructiveness of volcanic eruptions:

1. Volcanoes emit about 130 million tons of CO_2 (carbon dioxide) each year. However, the evidence reveals that volcanic eruptions were much greater in the past. A logical question would be, "How does evolution take place in such a destructive environment?"
2. About 90% of the gases emitted in volcanic eruptions consist of water vapor in the form of steam. Geologists claim that this water vapor helped form our oceans. It would take a lot of volcanic eruptions to create oceans! But the catch-22 is that the more volcanic eruptions, the more poisonous gases. The amount of CO_2 emitted because of these eruptions would have also produced a toxic atmosphere.
3. Oxygen is emitted from volcanoes but oxygen also *destroys developing DNA*. This is a conundrum, because life requires oxygen in a very careful balance. Therefore, all life must have been originally created already able to utilize oxygen. And so, evolution can not explain how life acquired the use of oxygen.
4. Volcanic eruptions emit hot ash, which is initially deadly. The ash would have been extremely dangerous to air-breathing animals. Mt. St. Helens alone, a small volcano, produced 540 million tons of ash in nine hours, spreading over 22,000 square miles. Multiply this by 50,000 (the estimated number of volcanoes in the past) and that would spell disaster to developing life. Remember, most of these volcanoes were erupting at the same time. Even secular scientists accept this, as evidenced by how they date the past eruptions.
5. Volcanic eruptions emit sulfur dioxide, currently at about 22 million tons per year. Multiply this those 50,000 eruptions, and that would be deadly to developing life. Not only this, but SO_2 (sulfur dioxide) combines with water to form sulfuric acid.
6. The fossil record indicates that the past environment of the earth was warm and tropical-like. Volcanic eruptions would have blocked the sun's radiation through the emission of ash, which in turn would have cooled the earth's temperature, significantly changing the earth's climate.
7. In addition, geologists insist on adopting the Plate Tectonics model for the evolving earth, all the while life is evolving and volcanoes are spewing their stuff. The pulling apart of the continents would have put additional strain on developing life.
8. If plates are moving apart at about one inch per year, in just 65 million years they would have moved over 1,000 miles since the age of the

dinosaurs, about 4000 miles since the Permian Period, enough to upset every ecosystem that has ever existed since the Permian. This is roughly the distance from Seattle, Washington to London, England. If plates are moving apart as geologists insist, *then volcanic activity would also likely be proportionately increased.* Could life thrive in this type of geologic environment? We have already seen the improbability of this scenario based on what we know about even one catastrophic volcanic eruption.

9. Volcanoes emit hydrogen sulfide, as well as the acidic gases hydrogen chloride and sulfur dioxide. These latter two hurt the eyes and throat: they are harmful to developing life. In addition, lakes and other bodies of water dissolve these gases and the water becomes acidic which would burn a swimmer's skin in minutes. Likewise, developing life in water would have to fight these destructive conditions.

10. Volcanoes emit hydrogen fluoride, a poisonous gas strong enough to etch glass!

In the evolutionary scheme of things the formula for the first evolving life would be

Time
+ self-organizing inorganic chemicals
+ volcanic eruptions
+ self-organizing organic molecules

　　LIFE

Evolutionists insist that the main creative force in the beginning was the volcanoes or fire. But this is contradictory to 2 Peter 3:5-7 which states,

> *For when they maintain this, it escapes their notice that by the word of God the heavens existed long ago and the earth was formed out of water and by water, through which the world at that time was destroyed, being flooded with water. But by His word the present heavens and earth are being reserved for fire, kept for the Day of Judgment and destruction of ungodly men.*

Fire is a destructive force, not a creative force. Water is the key to life, and is recorded so, in Genesis.

If volcanoes were not a part of the original creation, then when did they appear? Genesis 7:11 might provide us with the framework:

> *In the six hundredth year of Noah's life, in the second month, on the seventeenth day of the month, on the same day all the fountains of the great deep burst open, and the floodgates of the sky were opened.*

So, how old are the Hawaiian Islands?
The origin of the Hawaiian Islands (volcanoes) would most likely have begun shortly before the end of the Flood or shortly after the Flood, by the ongoing tectonic activity brought on by the initial breaking up the fountains of the great deep. They would not have been a part of the original creation.

Thought Questions

1. Briefly describe the pH Scale. What number represents the most acidic substances?

2. Name the products of volcanic eruptions and volcanic flows.

3. What does the word pyroclastic mean? In what ways is it destructive to Earth?

4. Why does it make more sense to place the appearance of volcanoes during and after the Flood?

Activity: Watch a video presentation of the recent lava flows on the Big Island. As you watch this video, write down the things that you see that are destructive results of the lava flows.

Lesson Four
Types of Volcanoes

Word Challenges: *andesite, caldera, cinder cone/volcano, composite volcano, dome volcano, fissure volcano, magma, rift, scoria, shield volcano, stratovolcano, scoria, viscosity, volcanic vent*

Not all volcanoes are alike. Learning the various features of the different volcanoes will help you identify different volcanic rocks in a later section. Here are a couple of helpful illustrations of the various types of volcanoes.

Fissure volcano, Hawaii

Shield volcano, Mauna Kea, Hawaii

Dome Volcano, Mt. St. Helens

Cinder volcano, Creaters of the Moon National Monument, Idaho

Stratavolcano or Composite volcano, Mt. Fuji, Japan

Caldera, Crater Lake, Oregon

Volcano Type	Characteristics	Examples	Simplified Diagram
Flood or Plateau Basalt	Very liquid lava; flows very widespread; emitted from fractures	Columbia River Plateau	
Shield Volcano	Liquid lava emitted from a central vent; large; sometimes has a collapse caldera	Larch Mountain, Mount Sylvania, Highland Butte, Hawaiian volcanoes	
Cinder Cone	Explosive liquid lava; small; emitted from a central vent; if continued long enough, may build up a shield volcano	Mount Tabor, Mount Zion, Chamberlain Hill, Pilot Butte, Lava Butte, Craters of the Moon	
Composite or Stratovolcano	More viscous lavas, much explosive (pyroclastic) debris; large, emitted from a central vent	Mount Baker, Mount Rainier, Mount St. Helens, Mount Hood, Mount Shasta	
Volcanic Dome	Very viscous lava; relatively small; can be explosive; commonly occurs adjacent to craters of composite volcanoes	Novarupta, Mount St. Helens Lava Dome, Mount Lassen, Shastina, Mono Craters	
Caldera	Very large composite volcano collapsed after an explosive period; frequently associated with plug domes	Crater Lake, Newberry, Kīlauea, Long Valley, Medicine Lake, Yellowstone	

Topinka, USGS/CVO, 1997, Modified from: Allen, 1975, Volcanoes of the Portland Area, Oregon, Ore-Bin, v.37, no.9

Fissure volcanoes

An example of a ***fissure volcano*** is this fissure that is part of Kilauea on the Big Island. This picture was taken during the Kamoamoa eruption in 2011.

Fissures actively propagating during Kamoamoa eruption, Kīlauea, Hawai'i: You can see lava just breaking the surface in the foreground crack.

Fissure volcanoes or vents are cracks in the earth through which the lava erupts, usually with non-violent eruptions. Typically, fissure vents are characterized by basalt lava flows. This type of volcano is usually hard to recognize from the ground and from outer space because it has no central *caldera* or opening and the surface is mostly flat. The volcano can usually be seen as a crack in the ground or on the ocean floor.

Shield Volcanoes
Shield volcanoes look like a huge shield, convex side up. They usually stretch for miles and are not easily visible close-up. Shield volcanoes are built almost entirely of fluid basalt lava flows. An example of a shield volcano is Mauna Loa on the Big Island of Hawaiʻi.

Mauna Loa, Hawaiʻi

Dome Volcanoes
An example of a *dome volcano* is Chaitén in the Gulf of Corcovado in southern Chile, South America.

Chaitén Volcano

Dome volcanoes often form inside of the craters of existing **composite volcanoes** or **stratovolcanoes**. An example is Mt. St. Helens. Notice, in the following picture, the new dome currently building within the destroyed cone of Mt. St. Helens over just a six-year period.

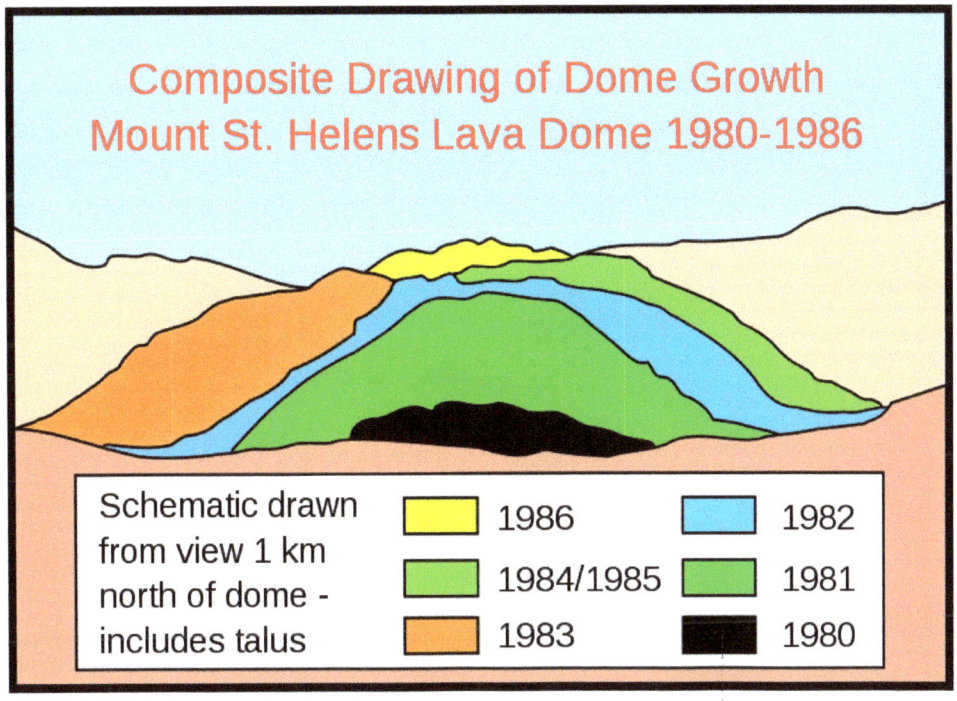

Ash or Cinder Cones

An example of an *ash/cinder cone volcano* is SP Crater in the San Francisco volcanic field of Arizona.

SP Crater, Arizona

Ash or cinder cones usually erupt just once in their life cycle. Another name for a cinder cone is *scoria cone* because the type of lava it erupts is the volcanic rock scoria, which is airy, lightweight, basaltic lava. They are characterized by emitting showers of hot cinders.

Another example of a cinder cone is one located on Mauna Kea shield volcano, Hawai'i.

Cinder cones located on Mauna Kea, which is a shield volcano

Composite Volcanoes or Stratovolcanoes

Two examples of a composite volcano (stratovolcano) are Mt. Fuji in Japan and Mt. Hood in Oregon.

Mt. Fuji, Japan

Mt. Hood, Oregon

Mt. St. Helens, Washington: Stratovolcanos are highly explosive.

Composite or stratovolcanoes build high cones made of repeated explosive volcanic eruptions and lava flows. These can build up very quickly.

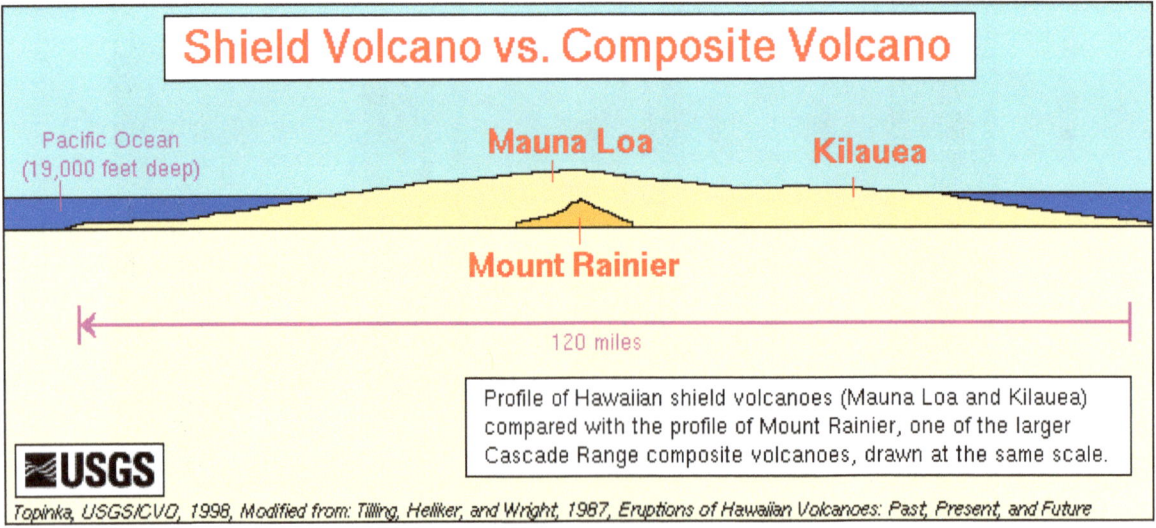

Mount Rainier is an example of a composite volcano or stratovolcano. It is the tallest stratovolcano in the continental US.

Caldera

A *caldera* is a huge eruption of ash, pyroclastic flows, and lavas that occupy a massive crater. An example of a ***caldera volcano (super volcano)*** is the Yellowstone Caldera in Wyoming. Its crater is estimated at 45 miles by 35 miles!

This picture is a satellite view of the Yellowstone Caldera. The shadowed area is the extent of the caldera. The large dark blue area is Yellowstone Lake, a glacial lake and part of the caldera. Calderas are so large that they are not easily noticed. It took scientists a couple of hundred years to recognize these depressions as calderas. The advent of aerial photography helped to see these immense catastrophic landforms. The Yellowstone Caldera is 45 miles long and 35 miles wide – truly a huge catastrophic and destructive volcano! What event could have triggered such huge eruptions in our past history? Were these calderas just part of the on-going recycling geologic process envisioned by James Hutton? Or were they part of the Genesis catastrophic flood that destroyed God's creation as a result of man's rebellion?

Laccoliths

What is a laccolith? The word comes from two Greek roots meaning, *pond stone*. A ***laccolith*** is technically a pool or pond of ***magma*** that did not quite erupt through the earth's surface and so, formed a hill or mound usually in sedimentary rock. In other words, the rise of magma never quite became a volcano. Laccoliths typically consist of relatively ***viscous*** magmas. The rock formed looks very much like *andesite* lava and gray *diorite* - intermediate rocks between the highly viscous lava, rhyolite and basalt lava.

> ***Viscosity*** is a measure of resistance to flow. The more quartz magma and lava contain, the more resistant to flow it is. In other words, if a lava is highly viscous, it moves very slowly. Rhyolite lava contains the most quartz so it is highly viscous. Lava that is highly viscous tends to be more explosive. Yellowstone is an example of a rhyolitic volcano or caldera, highly viscous lava. This is what made it tremendously explosive. The volcanoes of Hawai'i are basaltic and therefore contain much less quartz. This makes them less viscous, and so less resistant to flow. Eruptions from these types of volcanoes are more mild, and non-explosive.

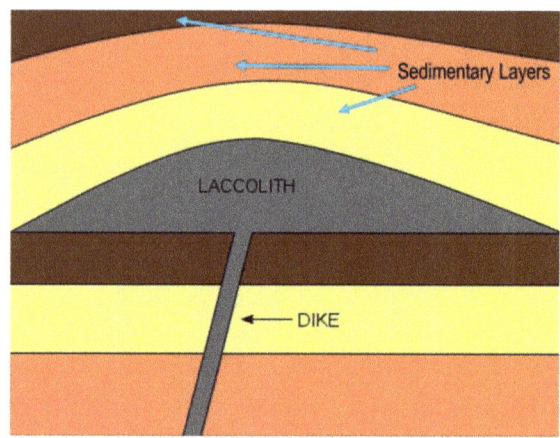

Diagram of a laccolith

Laccoliths are usually inconspicuous, as they are generally covered over by grass, shrubs and trees and could be mistaken for just an odd hill. A person can drive by laccoliths and not even be aware of them. But, once you recognize one, you will never forget it.

The following three pictures are of laccoliths. The erosion (or planing) would have been rapid massive sheet erosion produced during the second stage of the Genesis Flood. The order of formation would likely be: (1) The Flood laid down the sedimentary layers of sediment. (2) Magma started to rise through the still pliable sediments. (3) The receding waters of the sheet flow phase of the Genesis Flood rapidly eroded the softer sedimentary layers, leaving the remnant of a volcanic intrusion – the laccolith.

A laccolith in central Montana

This laccolith is part of the La Sal Mountains in Southeastern Utah. It is located in a sandstone canyon a few miles outside of Moab, Utah. It is a perfect textbook illustration of an intrusion within sedimentary rock, having never quite erupted through, with the softer sandstone washed away during the channelized erosion stage of the Flood.

Square Butte northwest of Great Falls, Montana is a laccolith with a planed surface (caused by the planing action during the sheet erosion phase of the Flood).

Some think that Devil's Tower in Wyoming is a laccolith. Others think it is simply a volcanic neck, the surrounding softer sediments having been eroded away. Of course, the big question there is whether the sediments were rapidly eroded away by receding Floodwaters or were slowly eroded away over millions of years. That is a question that cannot be answered by science, but only by one's worldview.

Devil's Tower, Wyoming

The La Sal Mountains – southeastern Utah

Remnants of the sedimentary layers originally laid down by the Flood: The La Sal Mountains, in southeastern Utah, are themselves huge laccoliths. You can get an idea of just how thick the sediments must have been, as the La Sals intruded into them, but did not quite erupt as volcanoes. It was at the 8,500-foot level of these mountains that well-preserved theropod dinosaur tracks are preserved in sandstone.

The Ring of Fire

The largest concentration of volcanoes in the world is in what we call The Ring of Fire. The Hawaiian Islands are a part of this. The Ring of Fire is an area where many earthquakes and volcanic eruptions occur in the basin of the Pacific Ocean. In a 25,000-square mile horseshoe-shaped arc from Australia to South America is located the most active volcanic activity chain in the world. The Ring of Fire has 452 volcanoes and is home to over 75% of the world's active and dormant volcanoes as well as thousands of earthquakes each year. About 90%

of the world's earthquakes and 81% of the world's largest earthquakes occur along the Ring of Fire. Take a look at this map of the Ring of Fire.

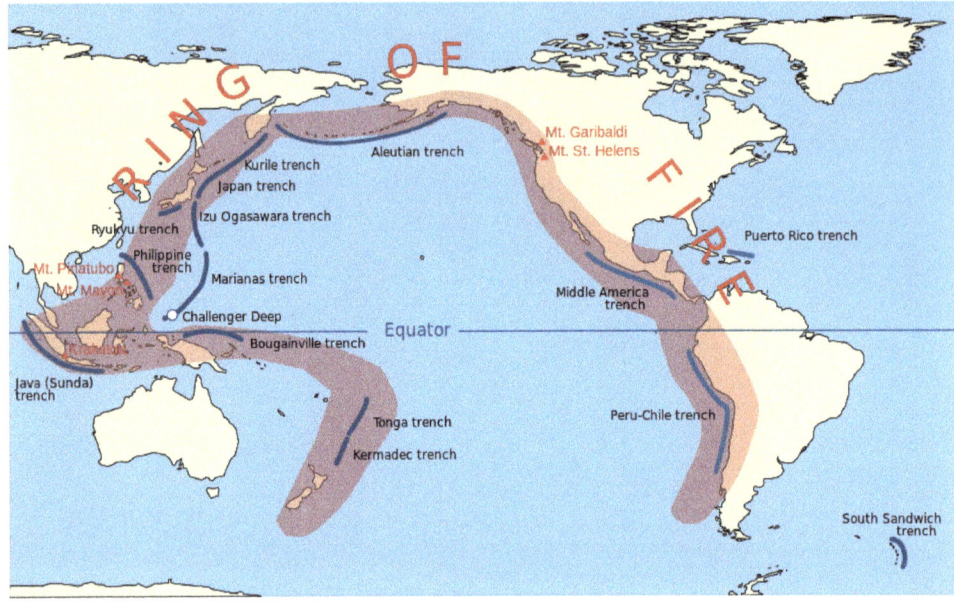

Take a look at the following map. This map is of the ocean floor. It was only well into the 20th Century that geologists could actually "see" the ocean bottom through special imaging technology. What they discovered amazed them and in an indirect way gave more credence to the Genesis account of the Flood. The ocean floor reveals cracks all around the globe. These huge cracks are what geologists call **oceanic trenches**. Some of these would dwarf the Grand Canyon in size and depth. These cracks likely would have been part of the breaking of the fountains of the great deep recorded in Genesis 7:11.

The volcanically active chains of earthquakes and volcanoes, such as the Ring of Fire, are in line with many of these cracks in the earth. (Incidentally, the Hawaiian Islands are not on line with these **rifts**. A rift is a crack in the earth out of which lava flows.) Modern secular geologists have divided these cracks into plates that they claim have been moving apart for hundreds of millions of years. The idea is called **plate tectonics**. (*Tectonics* is from a Greek word, meaning, *building*.) They believe that the continents are part of these plates which have shifted and moved around the globe. But it is only an idea. It has never been observed or scientifically verified. Earth movement has been studied and recorded, but that is a different study than continental movement. We will talk a bit more about plate tectonics in Lesson 7.

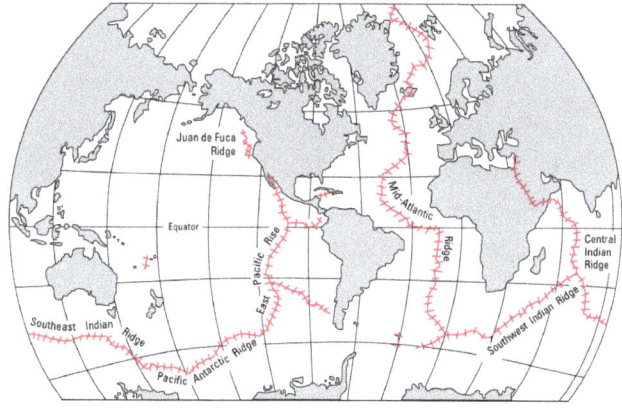

A Biblical way to look at this map is simply to observe what seems to be the remnant of what happened when the fountains of the great deep burst open. The earth, including what is now the ocean floor, is now scarred and altered by a massive flood catastrophe, the likes of which will not happen again. The earthquakes, volcanoes, climate changes and extremes are simply the vestiges of a violent time in Earth's recent past through which God judged an evil world.

Parts of a Volcano

Although the volcanoes are different from one to another, they all share some common parts. Study this diagram of a typical volcano and learn the various parts.

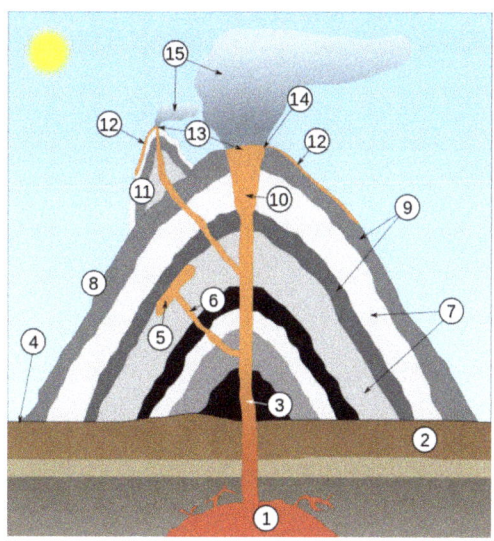

1. Magma chamber 2. Bedrock 3. Conduit (pipe) 4. Base 5. Sill 6. Branch pipe or dike 7. Layers of ash emitted by the volcano (layers of ash and **lava** do not necessarily alternate like this) 8. Flank 9. Layers of **lava** emitted by the volcano 10. Throat 11. Side vent or Parasitic cone 12. **Lava flow** 13. Vent 14. Crater 15. Ash cloud

Thought Questions:

1. Name the six types of volcanoes.

2. What is the difference between the lavas of lava flows and the lava produced by stratovolcanoes?

3. The idea that the earth is divided into plates that have moved away from each other through time is called what?

4. The area which includes most of the earth's volcanoes and earthquakes is called what?

5. What specific major historical geologic event took place in the 600th year of Noah's life on the first day of the Genesis Flood?

6. What is a laccolith?

Activity:
Experiment with the idea of viscosity. Find different things to test that are pourable or moldable, like ketchup, corn syrup, chocolate syrup, oil, honey, mayonnaise, peanut butter, etc. You will need a cookie sheet or something similar that is flat.

1. Measure out one tablespoon of one of the foods that you gathered.
2. Place a tablespoon of each at the top of a cookie sheet.
3. Tilt the cookie sheet. Do this for 5 seconds, then lay cookie sheet flat.
4. Which moved the most? Which least? Which has the highest viscosity? The lowest?

Thought question: Why is lava that is highly viscous more explosive?

Activity: Supply the missing labels for the volcano parts.

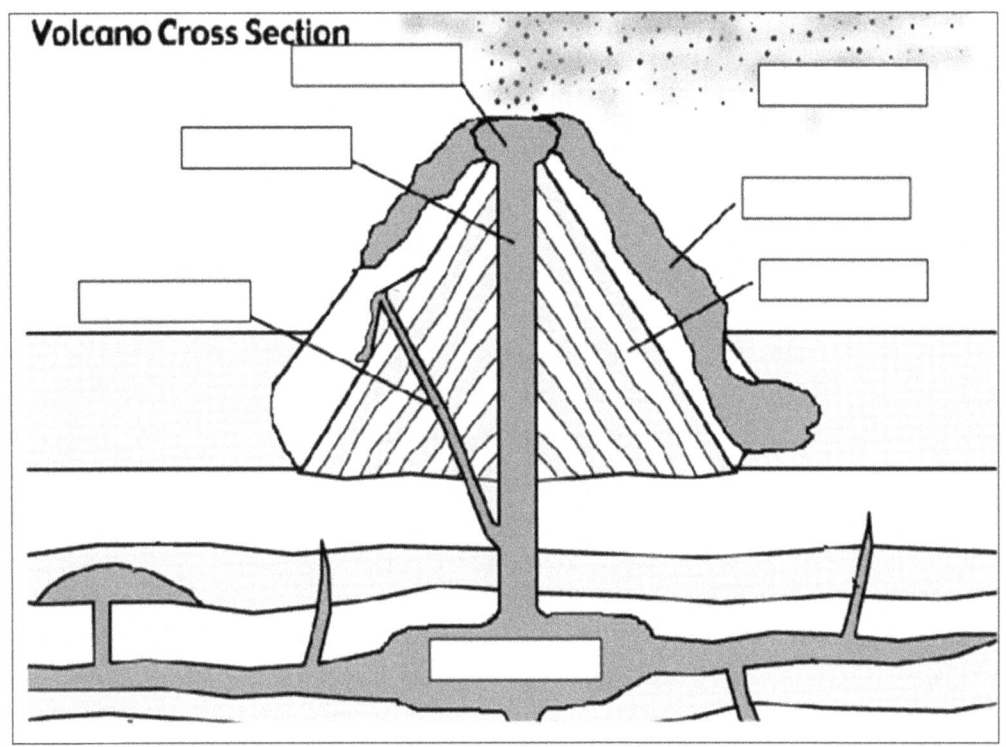

Lesson Five
Types of Eruptions

Word Challenges: *andesitic, fissure, lapilli, magmatic eruption, phreatic eruption, phreatomagmatic eruption, rhyolitic*

During a volcanic eruption, lava, tephra (ash, *lapilli*, volcanic bombs and blocks), and various gases are expelled from a **volcanic vent** or fissure. What kinds of factors influence the type of volcanic eruption that will occur? Volcanologists have identified three different factors:

1. *Superheated steam and gas is* released under pressure producing **magmatic eruptions** characteristic of Hawaiian volcanoes, more on the order of basaltic showers rather than explosive eruptions like Mt. St. Helens.
2. Thermal contraction from chilling on contact with water or from steam explosions causing **phreatomagmatic eruptions.** A phreatomagmatic eruption is the result of an explosive *water-magma interaction*, characteristic of the eruption of Mt. Usu in Japan in April 2000. Mount Usu formed on the southern rim of the caldera containing the lake, which in turn came into contact with the hot magma. Large amounts of steam and magmatic gases were emitted.
3. The ejection of volcanic rock from past eruptions that have blocked the conduit or pipe of a volcano and have thus been under great amounts of pressure causing **phreatic eruptions.** A phreatic eruption, also called a phreatic explosion or ultravulcanian eruption, occurs when *magma heats ground or surface water.* The extreme temperature of the magma anywhere from 932 to 2,138 °F causes near-instantaneous evaporation to steam, resulting in an explosion of steam, water, ash, rock, and volcanic bombs. At Mount St. Helens, hundreds of steam explosions preceded the 1980 explosive eruption.

All three of these factors involve the interaction of magma and water to some degree or another. Is it any wonder that during the great Genesis Flood many violent volcanoes would have been active and catastrophic such as the Yellowstone Caldera?

Types of eruptions have been classified and named after historically significant volcanic eruptions.

- **Hawaiian eruptions** – These are chacterized by the non-explosive eruption of very fluid basaltic lava. The amount of lava extruded from these ***Hawaiian eruptions*** is less than half of that found in any other volcanic type. The continuous eruption of small amounts of lava builds up the large, broad form of the shield volcano. Eruptions do not necessarily take place at the same central vent, like other volcanoes. Eruptions often occur at vents scattered around the shield volcano, and at fissure vents coming from the center.

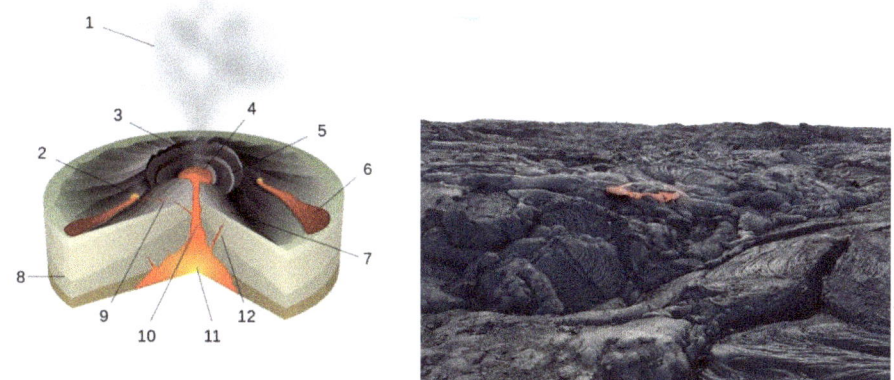

Hawaiian Eruption and pahoehoe lava from Kilauea
Eruption key: 1. Ash plume 2. Lava fountain 3. Crater 4. Lava lake 5. Fumaroles 6. Lava flow 7. Layers of lava and ash 8. Stratum 9. Sill 10. Magma conduit 11. Magma chamber 12. Dike.

- **Strombolian eruptions** are named after the volcano Stromboli. These eruptions are characterized by high gas content and blasts of lapilli and volcanic bombs. This combination makes these eruptions louder. They also do not typically produce things like Pele's tears and Pele's hair. They are not characterized by sustained lava flows.

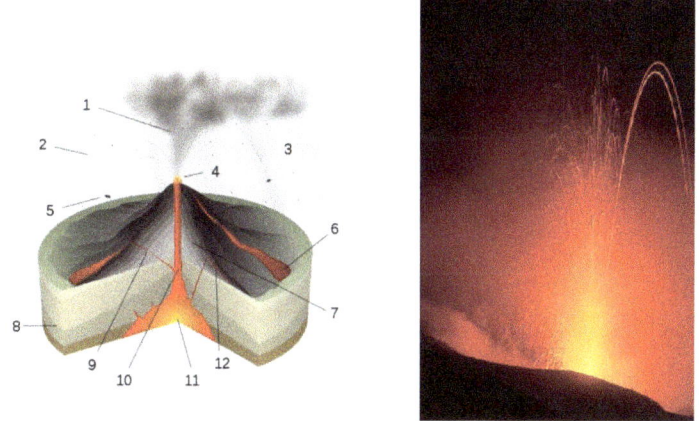

Diagram of a Strombolian Eruption and Mt. Stromboli, Italy
Eruption key: 1. Ash plume 2. Lapilli 3. Volcanic ash rain 4. Lava fountain 5. Volcanic bomb 6. Lava flow 7. Layers of lava and ash 8. Stratum 9. Dike 10. Magma conduit 11. Magma chamber 12. Sill

- **Vulcanian eruptions** are characterized by highly viscous lava eruptions. They are therefore characterized by explosive, gaseous lavas. Vulcanian deposits are andesitic to dacitic rather than basaltic. Andesite and dacite belong to the intermediate class of lavas which are gray to brown in color.

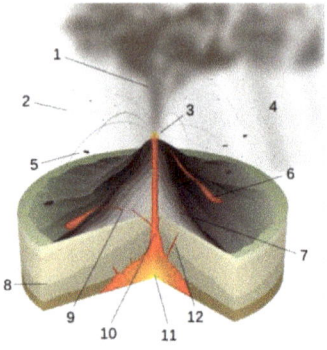 

Diagram of Vulcanian Eruption and Irazu Volcano in Costa Rica
Eruption key: 1. Ash plume 2. Lapilli 3. Lava fountain 4. Volcanic ash rain 5. Volcanic bomb 6. Lava flow 7. Layers of lava and ash 8. Stratum 9. Sill 10. Magma conduit 11. Magma chamber 12. Dike

- **Peléan eruptions are** named after the volcano Mount Pelée in Martinique, the site of a massive Peléan eruption in 1902. This was one of the worst natural disasters in history. Pelean eruptions are characterized by huge amounts of gas, dust, ash and lava fragments. Their eruptions are highly explosive and blown out of the central crater of the volcano. The eruption is driven by the collapse of a massive dome. The type of volcanic rock formed in this type of eruption is rhyolite, dacite, and andesite. These lavas are higher in quartz than the basaltic lavas.

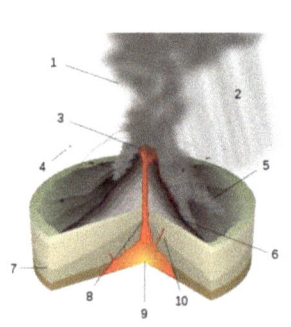 

Diagram of a Pelean Eruption and Mayon Volcano, Philippines, 1984
Eruption key: 1. Ash plume 2. Volcanic ash rain 3. Lava dome 4. Volcanic bomb 5. Pyroclastic flow 6. Layers of lava and ash 7. Stratum 8. Magma conduit 9. Magma chamber 10. Dike

- **Plinian eruptions (or Vesuvian)** – Plinian eruptions are also known as Vesuvian eruptions, and are named for Mount Vesuvius, after the famous eruption of AD 79. The name Plinian comes from the Roman historian Pliny the Younger who recorded the disaster at Pompeii. Plinian eruptions start in the magma chamber. Gases are stored in the magma resulting in massive eruptive columns of ash and pyroclastic material. These volcanic columns are the primary characteristic of Plinian eruptions. The ash clouds are ejected upwards of 28 miles into the atmosphere.

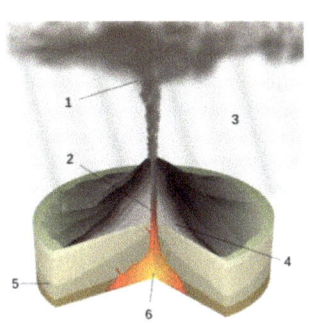

Diagram of a Plinian Eruption and the April 21st, 1990 eruptive column from Redoubt Volcano in Alaska, as viewed to the west from the Kenai Peninsula
Eruption key: 1. Ash plume 2. Magma conduit
3. Volcanic ash rain 4. Layers of lava and ash 5. Stratum 6. Magma chamber

- **Surtseyan eruption (or hydrovolcanic) or Phreatomagmatic eruptions** – These are eruptions that arise from interactions between water and magma – a type of volcanic eruption caused by shallow-water interactions between water and lava.

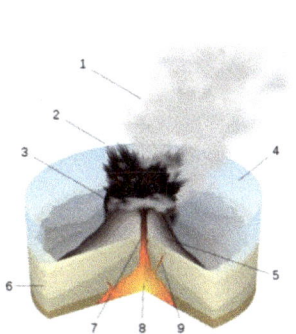

Diagram of a Surtseyan Eruption and the Island of Surtsey, erupting 13 days after breaching the water: a tuff ring surrounds the vent.
Eruption key: 1. Water vapor cloud 2. Compressed ash 3. Crater 4. Water 5. Layers of lava and ash 6. Stratum 7. Magma conduit 8. Magma chamber 9. Dike

- **Submarine eruptions** – occur underwater. It is estimated that 75% of all volcanic eruputive volumn comes from submarine eruptions near the mid-ocean ridges. Many of these submarine volcanoes have flat tops. These are called table mounts or seamounts. These are extremely interesting in that they are essentially flat-topped underwater volcanoes. They indicate that these were at one time above water. In our Biblical framework, flat-top volcanoes would be a reminder of the shearing effect of the sheet erosion phase of the Flood.

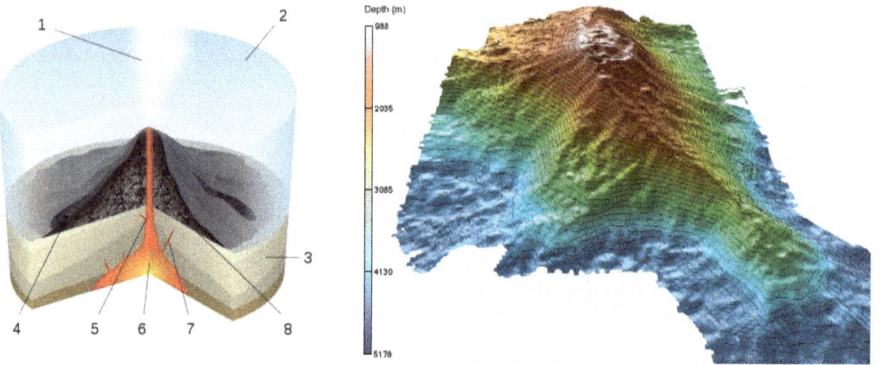

Diagram of a Submarine Eruption and Lō'ihi, a seamount, or underwater volcano, on the flank of Mauna Loa, the earth's largest shield volcano
Eruption key: 1. Water vapor cloud 2. Water 3. Stratum 4. Lava flow 5. Magma conduit 6. Magma chamber 7. Dike 8. Pillow lava

- **Phreatic eruptions (or steam-blast eruptions)** - Phreatic eruptions are characterized by hot steam explosions, resulting from erupting lava, inundated by water.

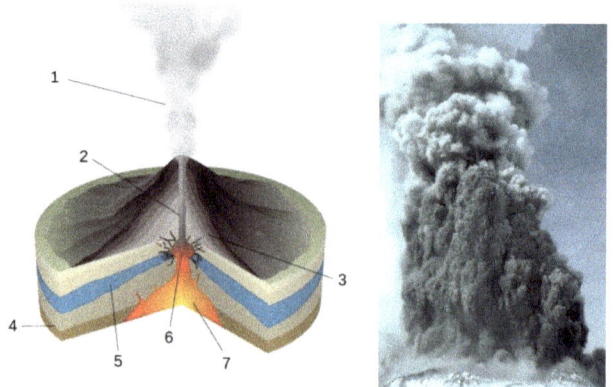

Diagram of a Phreatic Eruption and the eruption of Mt. St. Helens, 1980
Eruption key: 1. Water vapor cloud 2. Magma conduit 3. Layers of lava and ash 4. Stratum 5. Water table 6. Explosion 7. Magma chamber

Let's compare the relative explosiveness of the different types of volcanoes. Notice that the volcanoes that typically have less viscous lava are the ones that are most like what is in the Hawaiian Islands – the Hawaiian Eruption types.

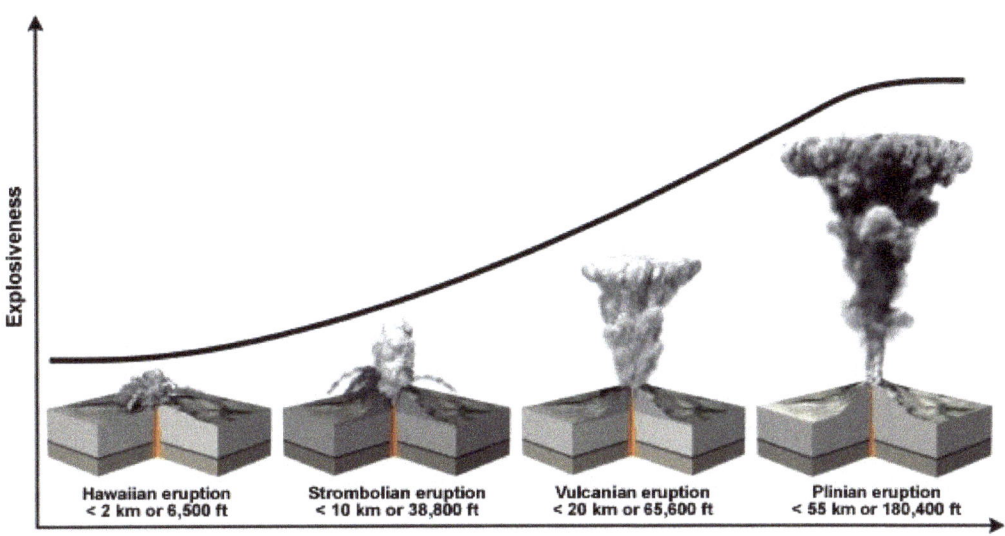

Thought Questions

1. The term tephra describes what kind of volcanic material?

2. What are the differences between a Hawaiian type eruption and a Plinian type eruption? Which one would be most likely to produce pyroclastic flows? Why?

3. What two common factors influence the explosivity of an eruption?

4. Which of the following eruptions is most likely to produce rhyolite lava? Hawaiian or Pelean? Why?

5. What is the earth's largest shield volcano?

6. What is a table mount?

7. What is a phreatic eruption?

8. Vulcanian eruptions are most likely to produce what kind of lava?

Activity: Volcanic Eruption Types
Make representations of the various eruption types. You can draw them or model them from clay, for instance. There is no need to label the parts, but be sure that your work accurately represents the particular kind of eruption you are illustrating.

Lesson Six
The Rocks and Minerals of the Hawaiian Islands

Word Challenges: *A'a', basalt porphyry, cinder,* dikes, intrusive, iron oxide, mafic, oxidation, *Pele's Tears,* phenocryst, *tachylite,* vesicular, *vitrophyre, pahoehoe,* veins, vitreous

The Rocks of Hawai'i
The Hawaiian Islands are of course, entirely volcanic, built from the volcanoes that make them up. The lava that comes from these volcanoes is basalt, which is black when fresh, and is high in the dark colored rock-forming minerals, olivine, pyroxene, biotite, amphibole, calcium feldspar, and iron.

Rock-forming Minerals
Study the pictures of the rock-forming minerals. Most of the rocks on Earth are made of the following minerals. Although other minerals are present in rocks, these are the predominant ones. You will notice that the rock-forming minerals are divided into six dark colored minerals and six light colored minerals. It is these minerals that give the rocks their dark and light colors. The variations in the colors of reds, greens, browns, etc., are due to *impurities*. These impurities are caused by other elements and minerals or chemical reactions. For example, the mineral serpentine produces the dark green color in the metamorphic rock, serpentinite. The chemical reaction between oxygen and iron produce the reddish to orange color in black basalt.

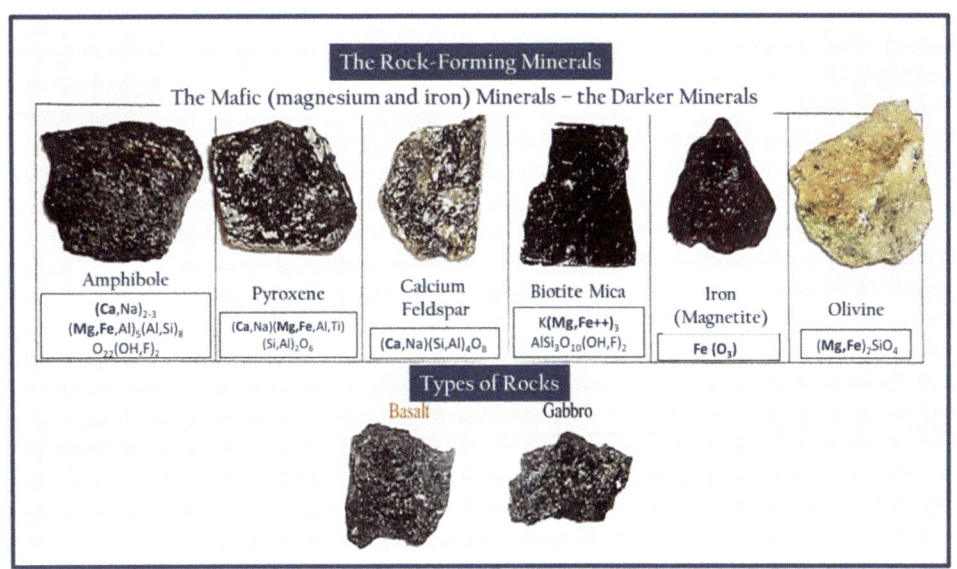

Dark-colored rock-forming minerals

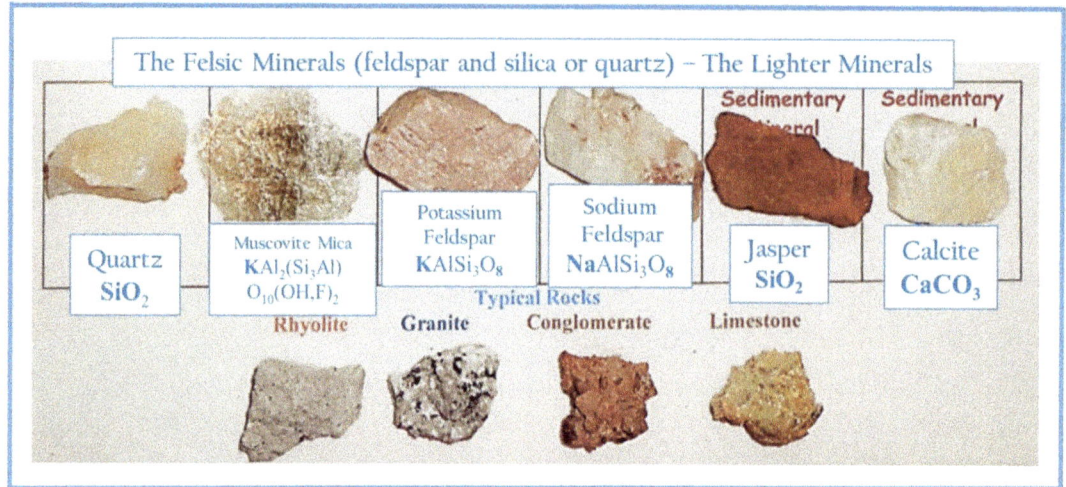

Light-colored rock-forming minerals

In addition, the darker rock-forming minerals and darker rocks are called **mafic** minerals and rocks. These five letters stand for magnesium and iron (*ma* from magnesium, *fic* being from the Latin word for iron). This word tells us that the rocks and minerals contain high proportions of magnesium in the form of olivine and iron usually in the form of magnetite or some other iron ore. The lighter-colored minerals and rocks are referred to as felsic. Because we are primarily dealing with the basalt lavas on the Hawaiian Islands, we won't address the *felsic* rocks and minerals. You can do that on your own.

Mineral Composition

But what are the minerals made of? The minerals are made of specially arranged patterns of elements. The earth is an amazing place! It is uniquely fitted for life. The elements and minerals are part of God's plan. That shouldn't surprise us. After all, "In the beginning God created the space and the earth".

So, let's talk about the most abundant elements in God's creation. The following chart tells us that out of all the elements (over 90 natural elements), there are just eight that form most of the earth's crust, and consequently, most of the rocks. These eight elements also make up the minerals, which form most of the rocks. On a very interesting note, if you change any one of these elements such that you decrease or increase the percentages, you will change the nature of our earth. God has given us just the right elements in the right percentages to make life possible. You will also notice that these elements are vital for our own health and are found in the foods we eat. Take a look at the following diagrams of the most abundant elements in the earth's crust and their chemical symbols. These diagrams will come in handy as you explore the world of rocks.

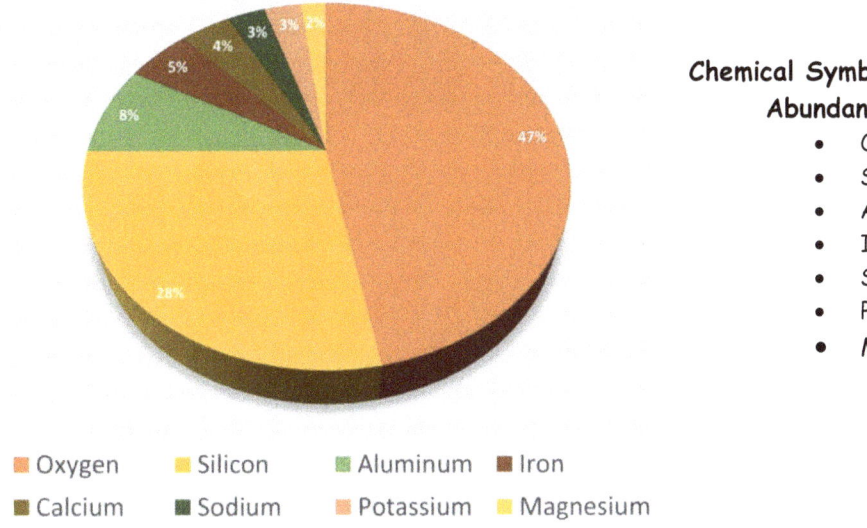

Volcanic Rocks

A volcanic rock is called a *fine-grained* rock in geology. That means that normally you cannot see the mineral makeup with the naked eye, only a general color. The colors are due to those rock-forming minerals that we talked about earlier.

Black basalt lava has a high concentration of the darker rock-forming minerals. It has very little quartz. When **oxidation** takes place in basalt, the result is an orange, red, brown and even yellow rock because of the high concentration of iron.

(Left) Black basalt lava (Right) Oxidized basalt

Vesicular basalt is lava that had high gas content as it was being formed. As the gas bubbles popped, vesicles formed in the lava.

Pahoehoe and A'a' lava

Pahoehoe is ropy shaped basalt lava. The Hawaiian word means, *smooth* lava. It also forms lava tubes. These surface features of pahoehoe seem to be due to the movement of very fluid lava under a congealing surface crust. Although not totally understood, with increasing distance from the source, pahoehoe flows may change into A'a' flows. This may be in response to heat loss, and consequent increase in viscosity, but we don't know for sure. A'a' is rough, spiky shaped lava. Both are prominent lavas in Hawai'i. The temperature of flowing basalt ranges from 1,292⁰ to 2,192⁰ F, depending on the distance from the vent.

Top row, left to right: vesicular basalt (oxidized because of the iron content), pahoehoe lava (recent basalt flow of Mauna Loa); middle row, left to right: basalt with olivine, A'a' basalt lava flow; Bottom row: the rock-forming mineral, olivine

Papakolea Beach (Green Sand Beach): olivine washes out of the basalt onto the beach

Other types of basalt include:

Aphanitic basalt (from the Greek word, *aphanes*, meaning, *unseen*) is basalt in which the mineral **phenocrysts** cannot be seen with the naked eye. A phenocryst is a crystal large enough to be seen with the naked eye.

Basalt porphyry is basalt that contains large, phenocrysts of sodium feldspar crystals.

Scoria is a form of highly vesicular basalt, extremely light-weight and airy. It is the basalt version of pumice that is normally seen with rhyolitic and andesitic eruptions.

Cinders are loose pyroclastic fragments, such as volcanic clinkers (loose, irregular fragments), cinders, volcanic ash, or scoria that has been built around a volcanic vent. They are very small and appear like black flakes of lava.

Aphanitic basalt

Basalt porphyry

Scoria   Cinder

Cinders (or clinkers)

Pele's Tears, named after the Hawaiian goddess of volcanoes, is a type of basaltic glass, but not glass as you and I think of it. The scientific name for basaltic glass is **tachylite**. Tachylite (also spelled tachylyte) is a ***vitreous*** form of basaltic volcanic glass. The color is a black or dark-brown, and it has a greasy-looking, resinous luster. It is very brittle and occurs in ***dikes, veins*** and ***intrusive masses.*** The word originates from the Greek word, *takhus* meaning *swift*.

Pele's Tears are so named for the tear-shaped small pieces of solidified lava drops formed when airborne particles of molten material fuse into tear-like drops of volcanic glass; Pele's Tears are jet black in color and are often found on one end of a strand of Pele's Hair. Pele's Tears is primarily a scientific term used by volcanologists.

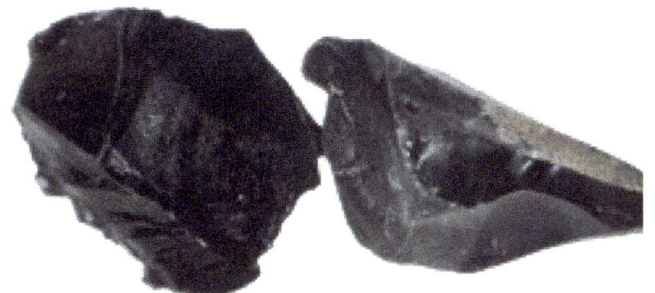
Tachylite – basaltic glass

Vitrophyre is a volcanic rock with larger crystals (phenocrysts) of sodium feldspar embedded in a glassy groundmass.

Vitrophyre

Black sand is really not sand in the traditional sense. Beach sand is made from small bits and pieces of quartz. Black sand is made from basalt that has come in contact with water when it was hot, and then shattered into tiny pieces. It was subsequently tumbled by the water.

(Left) Black sand from beach or, as it is most often called, Black Sand Beach
(Right) Punaluʻu beach (Black Sand Beach) located on the Big Island

Basalt bombs – these are pieces of basalt lava that are hurled through the air from basalt eruptions. As these pieces of lava hurl through the open air, they twist into many different interesting shapes. By the time they hit the ground, they have cooled enough to harden into those shapes. Bombs are also called *tephra.*

Ribbon bombs

Spindle bomb

Thought Questions

1. Name the mafic volcanic minerals. Name the felsic volcanic minerals.

2. Write out the chemical symbols of the eight most abundant elements

3. Name five examples of mafic volcanic rocks.

4. What is the word for *fire broken*?

5. A volcanic rock is referred to as what kind of rock?

6. Basalt eruptions tend to be what kind of eruption?

Activity
1. Take out the volcanic rocks and minerals that came with your kit. Become familiar with them so that you will be able to see the relationship between them and the rocks you will be looking at outdoors.

2. Select one famous volcano on one of the Islands and write a report about it. Include what is known about its history, its effects, and the type of rocks it erupted.

Lesson Seven
The Volcanoes, Earthquakes and Land Forms of Hawai'i

Word Challenges: *alluvium, bathymetry, hot spot, tuff cone, palagonite, tectonic earthquakes, volcano earthquakes*

The Hawaiian volcanoes include a very long chain of both active and inactive volcanoes, many of which are submerged. This entire chain is called the Emperor Seamount Chain.

The Emperor Seamount Chain is a mostly undersea mountain range in the Pacific Ocean that reaches above sea level in Hawai'i. It is composed of the Hawaiian ridge, consisting of the islands of the Hawaiian chain northwest to Kure Atoll, and the Emperor Seamounts. Together they form a vast underwater mountain region of islands and intervening seamounts, atolls, shallows, banks and reefs along a line trending southeast to northwest beneath the northern Pacific Ocean. The seamount chain, containing over 80 identified undersea volcanoes, stretches over 3,600 miles from the Aleutian Trench in the far northwest Pacific to the Lō'ihi seamount.

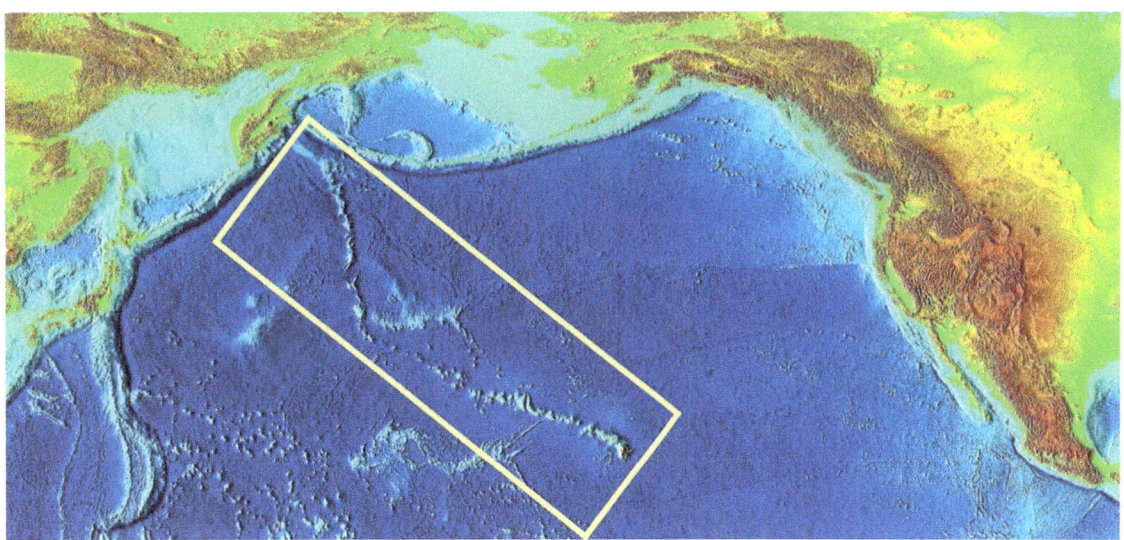

The Emperor-Hawaiian Seamount Chain

The habitable portion of the Emperor-Hawaiian Seamount Chain. The Big Island of Hawai'i is located to the far-right side of the satellite image.

The Big Island of Hawai'i

The Big Island is a composite of five shield volcanoes, Kohala, Mauna Kea, Hualālai, Mauna Loa and Kilauea. The Hawaiian word *mauna* means mountain. Driving over the center of the Island is quite an interesting adventure, as the trip crosses several lava flows where the highway has been destroyed and rebuilt several times.

The Hawaiian Island Chain of volcanoes, called an archipelago – satellite view

The Big Island of Hawai'i is made of five shield volcanoes.

Kohala, a shield volcano

The shield volcano, Mauna Kea

Hualālai, shield volcano

Hawaiʻi Volcanoes National Park is located on the Big Island of Hawaiʻi in the southeast corner of the Island. It was established as a national park in 1916. It includes an area of 505 square miles of diverse environments that include tropical forests, active volcanic lava flows, and even an arid climate in barren Kaʻū Desert. The climate of the Big Island is extremely varied, going from tropical sea level to the frigid summit of the earth's most massive active volcano, Mauna Loa, at 13,677 feet.

The Big Island of Hawaiʻi showing the location of the Park

The Park is made of two active volcanoes: Mauna Loa, the world's most massive shield volcano, and Kilauea, one of the world's most active volcanoes. Over half of the Park is designated as Hawaiʻi Volcanoes Wilderness. Active eruptive sites include the main caldera of Kīlauea and a more active but remote volcanic vent called Puʻu ʻŌʻō.

Mauna Loa

Mauna Loa is a shield volcano.

Mauna Loa, shield volcano: Hualālai is visible in the background.

This map shows the recorded historic lava flows of Mauna Loa.

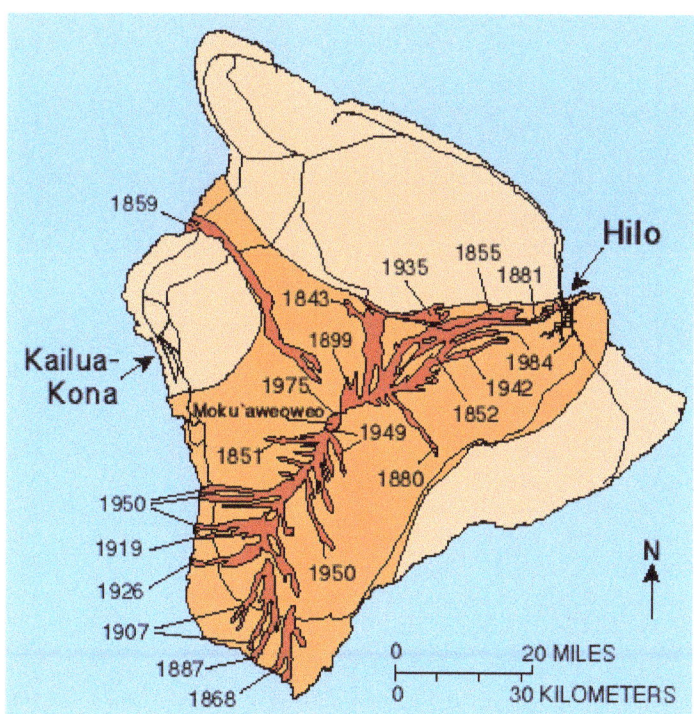

Mauna Loa is located in the center of the Big Island with historic lava flows branching out in several directions.

Lava erupts from a fissure high on the northeast rift zone of Mauna Loa at about 10,000 feet on the morning of March 25, 1984. Fountains were 30 to 150 feet high along the fissure, and lava output was estimated at 70,629,333 cubic feet per hour

Kilauea

Although separate volcanoes, Kilauea and Mauna Loa sit almost on top of each other. Kilauea, in the southeast corner of the Big Island, is a shield volcano on the flank of Mauna Loa. It also has a caldera near its summit.

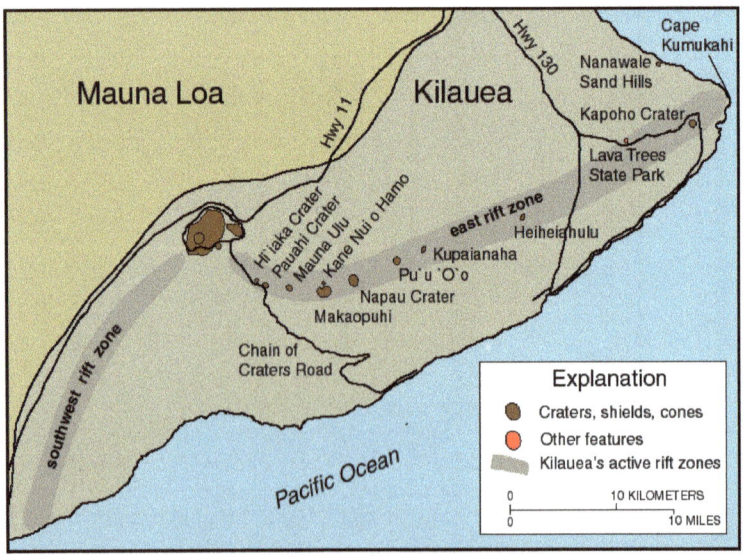

Mauna Loa and Kilauea: pictured are the various craters and cones that have opened up due to recent volcanic activity of Mauna Loa.

Kilauea, shield volcano; Mauna Loa can be seen in the background

Lava flows from Kilauea entered the community of Kalapana, located in the southeast part of the Big Island, southeast of Kilauea, briefly in November 1986. Lavas continued to erupt and then moved through the entire community in 1990.

The community of Kalapanna in June of 1990

Puʻu ʻŌʻō
Another volcanic cone located on the flanks of Mauna Loa is Puʻu ʻŌʻō.

It is the main vent of the longest eruption from the east rift zone in at least the past 1,000 years. It is now the largest cone on Kilauea, itself a caldera on Mauna Loa. Within 3 years of its first eruption in 1983, Puʻu ʻŌʻō reached a maximum height of 765 feet above the former ground surface. Beginning in 1993, **collapse pits** [depressions formed by a sinking or collapse of the surface that lies above a void or empty chamber, rather than from the eruption of a volcano or lava vent] began forming on the west flank. In early 1997, the entire west flank of the cone collapsed to form a large gap and a crater 630 feet deep. Collapses continue to change the shape of Puʻu ʻŌʻō. (From the most up-to-date report on the volcanic activity of Kilauea from the USGS.)

Puʻu ʻŌʻō Crater

Typical of basaltic eruptions, black clouds of cooled lava fragments (shown to the right of the lava shower) contain Pele's hair and Pele's tears, cinders, and pumice. Most of the cooled airborne material falls to earth close to the vent and contributes to the growing cinder-and-spatter cone. Pele's hair, the lightest material, can be carried downwind for many miles. A spatter cone is a low-sided hill cone made up of welded lava fragments called spatter. These are technically a little different from cinder cones, which are steep-sided cones built of loose pyroclastic fragments like cinders, scoria and ash. Cinder cone eruptions are typically gas-charged and more explosive than spatter cones.

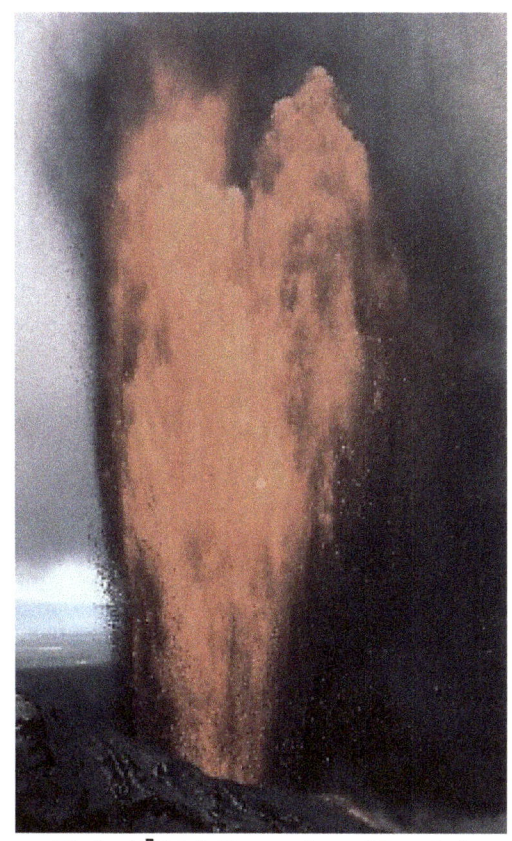

Puʻu ʻŌʻō full eruption of basalt lava

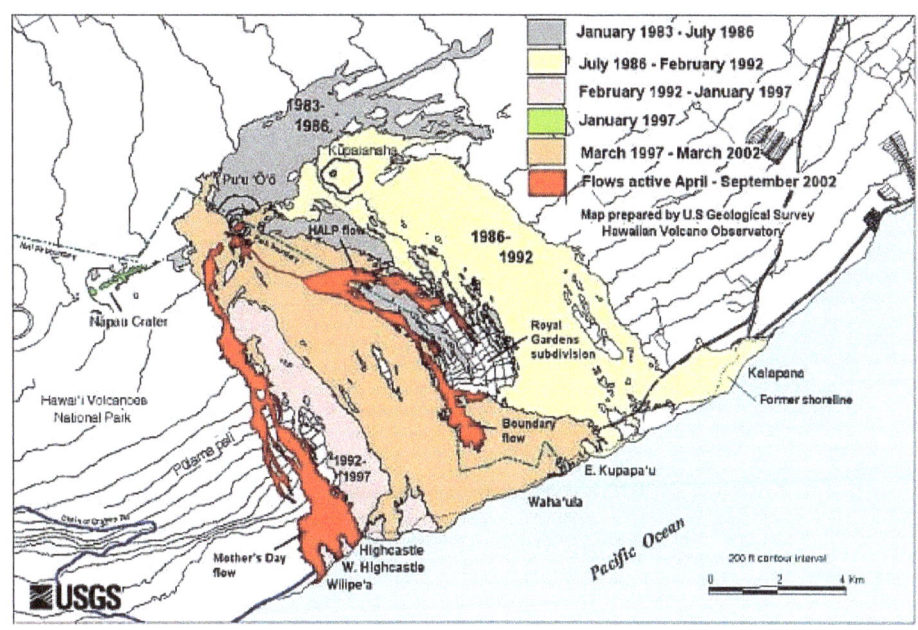

Volcanic activity of Puʻu ʻŌʻō from 1985 to 2002

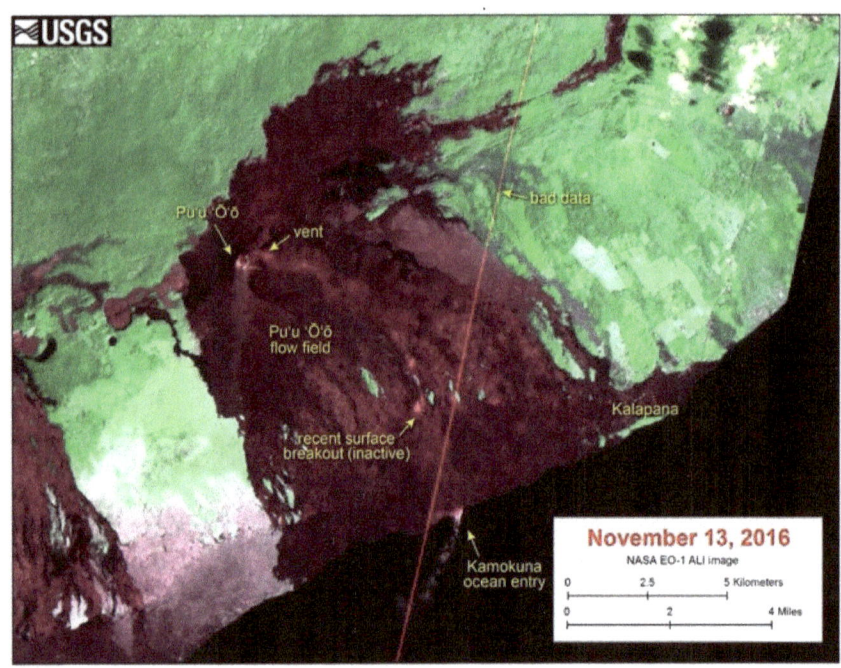

Satellite image of the Puʻu ʻŌʻō volcanic field in the southeast corner of the Big Island

Lava from the Puʻu ʻŌʻō cinder cone has flowed 14 miles into the town of Pāhoa. The lava has breached the boundary of the Pāhoa Transfer Station. Notice how the lava has pushed through the fence.

Lava flow over road from Puʻu ʻŌʻō

March 2015. Looking south toward the town of Pāhoa, and southwest toward the lava flow from Puʻu ʻŌʻō, outlined by burnt vegetation (lava flow is shown in the yellow box)

Lōʻihi volcano – Lōʻihi is a volcano located on the flank of Mauna Loa, off the coast of Hawaiʻi. Lōʻihi is underwater and technically called a ***seamount***. Seamounts are underwater volcanoes that can be active or inactive. The top of the Lōʻihi seamount is about 3,000 feet below sea level. Lōʻihi, meaning *long* in Hawaiian, is the newest volcano in the Hawaiian-Emperor seamount chain. At its summit, Lōʻihi Seamount stands more than 10,000 feet above the seafloor, making it taller than Mount St. Helens was before its catastrophic 1980 eruption. A diverse microbial community resides around Lōʻihi's many hydrothermal vents.

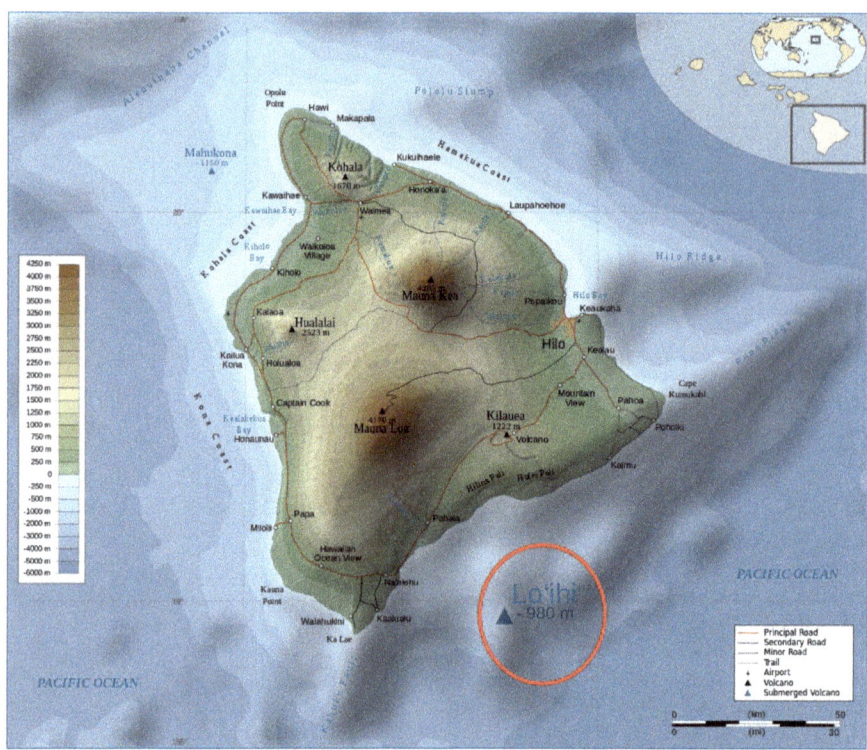
The seamount (or submarine volcano), Lōʻihi

Maui
Haleakalā National Park is on the southeastern side of Maui. The Park covers an area of 52 square miles of which 30 square miles is a wilderness area. It was originally created as part of the Hawaiʻi National Park along with the volcanoes of Mauna Loa and Kilauea on the Big Island in 1916. Hawaiʻi Volcanoes National Park was made into a separate national park in 1961. The main feature of Haleakalā National Park is the famous Haleakalā Crater. It is 6.99 miles across, two miles wide, and about 2,600 feet deep! Numerous volcanic features, including large cinder cones, dot the interior of the crater.

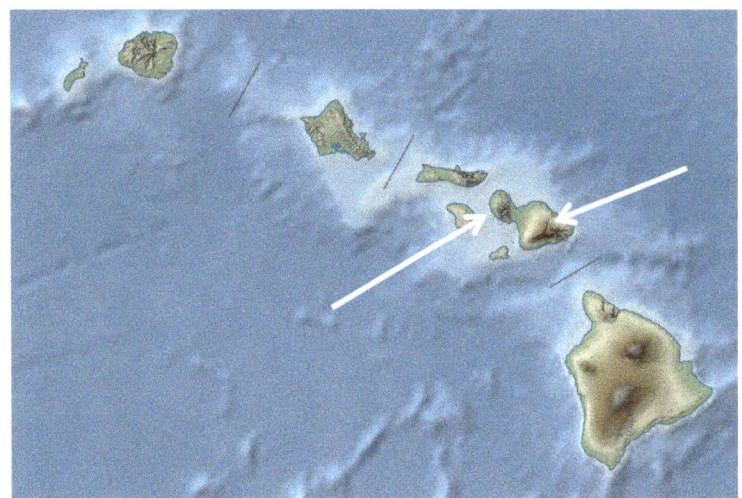

Haleakalā or the East Maui Volcano, is a massive shield volcano that forms more than 75% of the Hawaiian Island of Maui. The western 25% of the island is formed by another volcano, Mauna Kahalawai, also referred to as the West Maui Mountains.

Haleakalā crater

Known to the Hawaiians as Maui Komohana and to geologists as Mauna Kahalawai, the volcano is a much-eroded shield volcano that constitutes the western one-quarter of Maui.

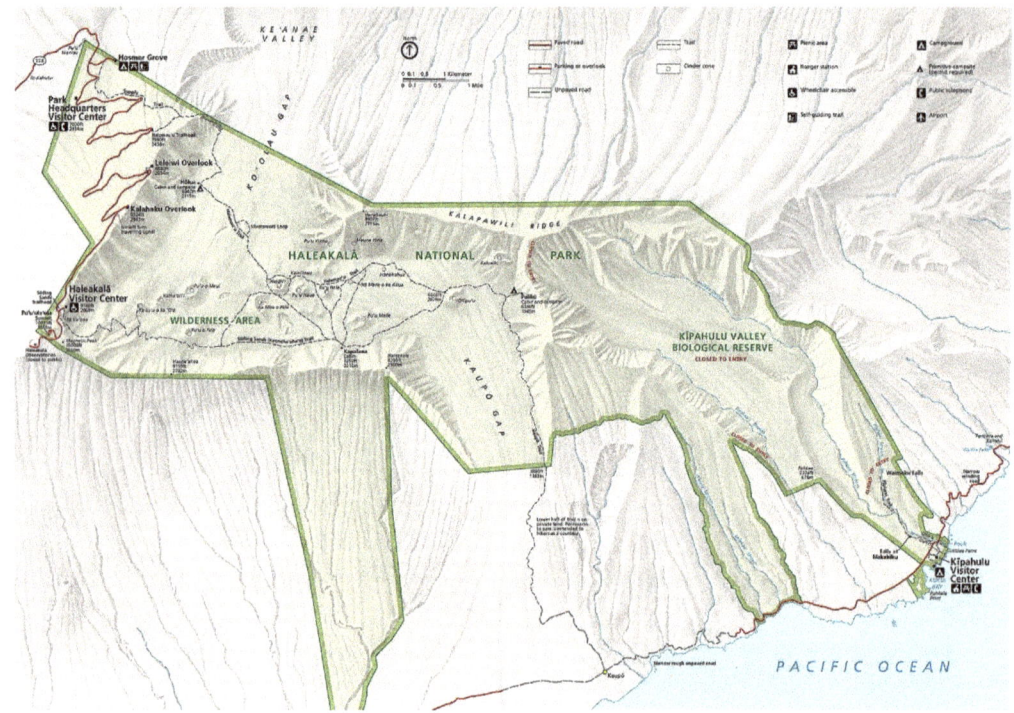

Map of Haleakalā National Park located on the island of Maui

There are mysterious rock features on Maui that require some explanation. Secular geologists have mapped these and have identified them as sedimentary rocks, sand dunes, mudflows and **alluvium**. Alluvium word (from the Latin, *alluere*, meaning, *to wash against*) is loose, unconsolidated (not cemented together into a solid rock) soil or sediments, which have been eroded, reshaped by water in some form, and redeposited in a non-marine setting. Alluvium is typically made up of a variety of materials, including fine particles of silt and clay and larger particles of sand and gravel. When this loose alluvial material is deposited or cemented into a lithological unit, or lithified, meaning to turn to stone, it is called an alluvial deposit.

Now, the mystery! The secular dates for these alluvial deposits have apparently been influenced by radiometric dates of the lavas around these deposits. Geologists acknowledge that the deposits are out of sync with the reported age of the lavas. The deposits are supposed to be older, but they are on top of the younger lava. The **Haleakalā** volcano is still active. Could it really be old on the order of millions of years? In addition, these deposits all appear to have been deposited in a marine setting, in other words, in lots of water. There are also reports that some of the lava flows are actually quite young and have been observed sometime between AD 1,400 and 1,800. What these deposits would indicate is an island that was most likely formed in the post-Flood period of the Genesis Flood, about 4,500 years ago and into the recent past.

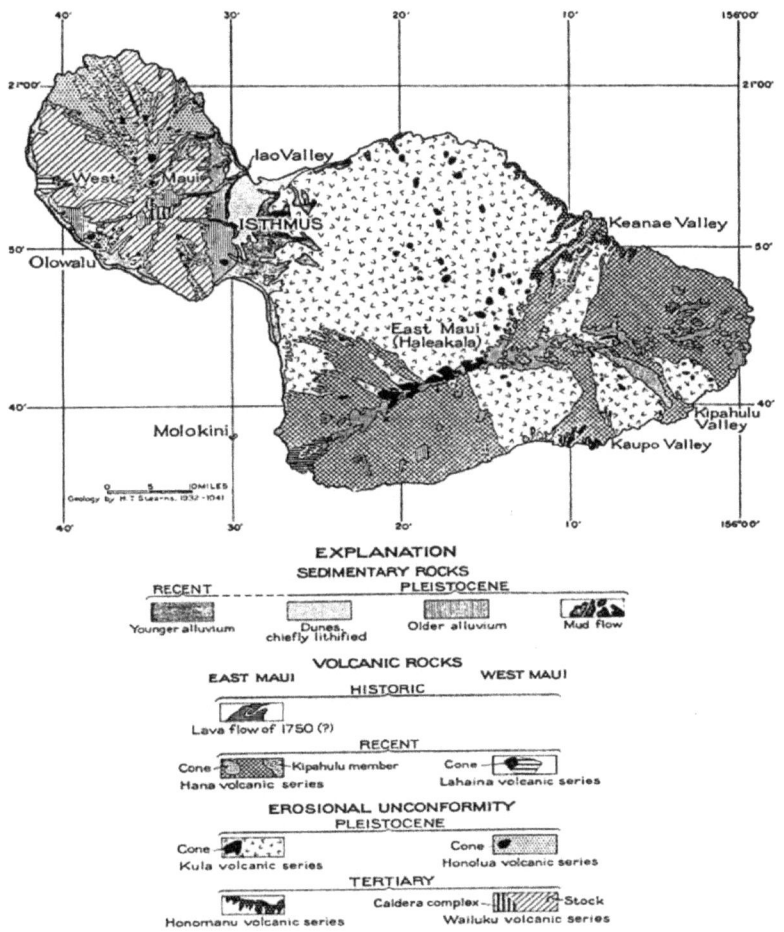

Geological map of the Maui deposit

Kahoʻolawe

Kahoʻolawe is the smallest of the eight main volcanic islands in the Hawaiian Archipelago. Kahoʻolawe is located about seven miles southwest of Maui and also southeast of Lānaʻi, with a total land area of 45 square miles. The highest point on Kahoʻolawe is the crater of Lua Makika at the summit of Puʻu Moaulanui, which is about 1,477 feet above sea level. Kahoʻolawe is a shield volcano.

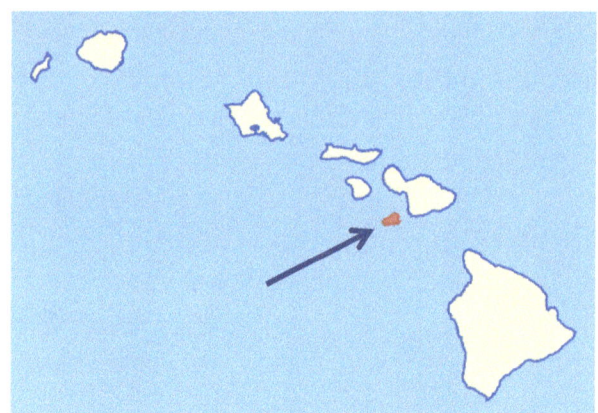
The tiny island of Kahoʻolawe in red

Landsat satellite image of Kahoʻolawe

Lānaʻi

Lānaʻi is the sixth largest of the Hawaiian Islands and the smallest publicly accessible inhabited island in the chain. It is also known as Pineapple Island because of its past as an island-wide pineapple plantation. As with all of the Hawaiian Islands, Lānaʻi is an undersea shield volcano, which means that most of its base is under water. Only the top is exposed, as is the case with the Big Island.

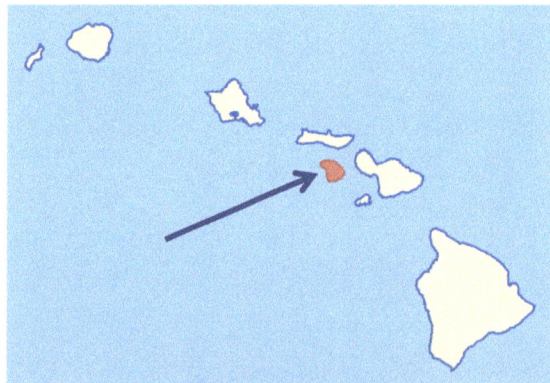
The island of Lānaʻi

Landsat satellite image of Lāna'i

Moloka'i

The East Moloka'i Volcano is an extinct shield volcano making up the eastern two-thirds of the island of Moloka'i. It is known as Wailau for the Wailau valley on the north side of the island.

The island of Moloka'i

Eastern Molokaʻi with a portion of Kamakou and Molokaʻi Forest Reserve: notice the deeply carved ravines on the elevated basalt slopes – evidence of a lot of water erosion and cutting.

Molokaʻi, south coast showing the past lava flow

Penguin Bank is the name given to a submerged shield volcano of the Hawaiian Islands. Penguin Bank lies immediately west of the island of Moloka'i, under relatively shallow water. The top of it is coral encrusted.

Bathymetry is the study **of the underwater depth of lake or ocean floors.** In other words, bathymetry is the underwater equivalent to topography, which is the study of surfaces on land. The name comes from Greek, *bathus*, meaning *deep*, and *metron*, meaning *measure*. Bathymetry is a fairly new field in geography.

Study the following map that was made using bathymetry. You can see where Penguin Bank lies, off the west coast of Moloka'i, southeast of O'ahu.

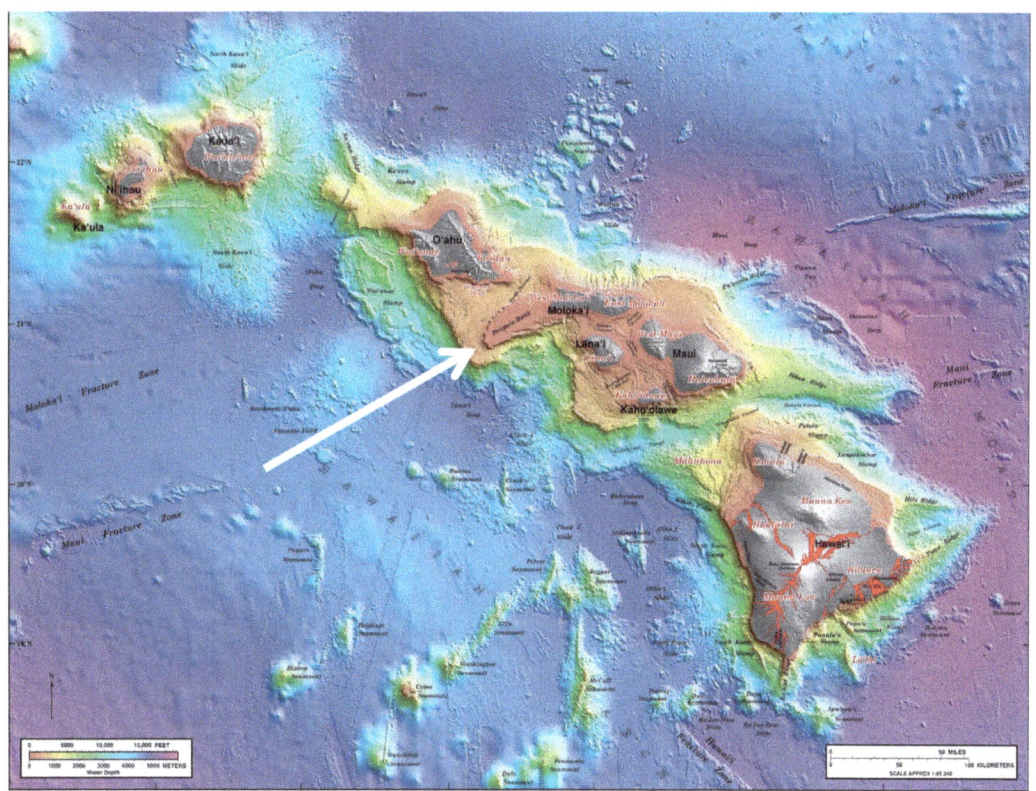

Bathymetric image of the Hawaiian Islands: The location of the volcano, Penguin Bank, is the light red area immediately west of Moloka'i

Oʻahu

Koʻolau Range is a name given to the fragmented remnant of the eastern, or windward. shield volcano of Oʻahu. It was designated a National Natural Landmark in 1972.

(Left) The island of Oʻahu   (Right) Deeply carved and eroded Koʻolau Range on the eastern side of Oʻahu

Oʻahu is home to several tuff cones. Tuff is a word used in volcanology to describe the fused product of super-heated ash, gases and bits and pieces of volcanic rock, cooled into a rock. A *tuff cone* is thought to be the result of hot water/steam eruptions. High rims characterize tuff cones. A tuff cone consists of thick-bedded pyroclastic flow deposits. Scoria-bomb beds derived from fallout are common. The tuffs composing a tuff cone have commonly been altered (***palagonitized***, from the rock named, *palagonite*), by either its interaction with groundwater or when it was deposited warm and wet. The pyroclastic deposits of tuff cones differ from the pyroclastic deposits of spatter cones by their lack of lava spatter, smaller grain-size, and excellent bedding. Typically, but not always, tuff cones lack associated lava flows.

Palagonite is a hydrothermally altered rock. It is a type of volcanic glass produced by the interaction of water and lava. It has a chemical composition similar to basalt (tachylite). Palagonite can also result from the interaction between water and basalt melt. The water flashes to steam on contact with the hot lava and the small fragments of lava react with the steam to form the light colored palagonite tuff cones common in areas of basaltic eruptions in contact with water.

The two most prominent tuff cones on Oʻahu are Diamond Head, and Koko Bay. Diamond Head is known to Hawaiians as Leʻahi. Its English name was given by British sailors in the 19th century, who thought that the calcite crystals on the adjacent beach were diamonds. Diamond Head is part of a broader series of volcanic cones and vents on Oʻahu, called the Honolulu Volcanic Series. There are five of these cones on Oʻahu.

Palagonite, altered volcanic rock (hydrothermally altered)

Diamond Head, a popular tourist site on the southeastern side of Oʻahu

Koko Bay crater on Oʻahu

Wai'anae Range (sometimes referred to as the Wai'anae Mountains) is the eroded remains of a shield volcano that makes up the western half of O'ahu. Its summit is 4,025 feet. These mountains also preserve the deeply cut ravines, all the way to the summit, that are so characteristic of the Hawaiian volcanic mountains. The water-cut ravines plus the apparently rapid severe erosion of very hard basalt rock speak of the relationship to a watery catastrophe of some kind.

Wai'anae Range, an eroded shield volcano on the west side of O'ahu

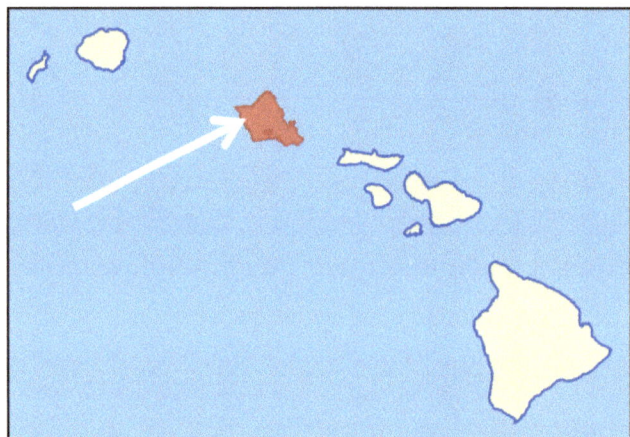
Wai'anae Range is on the west side of O'ahu

Ka'ena Ridge, also known as the Ka'ena Volcano, is a submerged remnant of a shield volcano that is to the north of O'ahu and once was part of the northern section of O'ahu. Ka'ena Ridge was one of three shield volcanoes to form O'ahu.

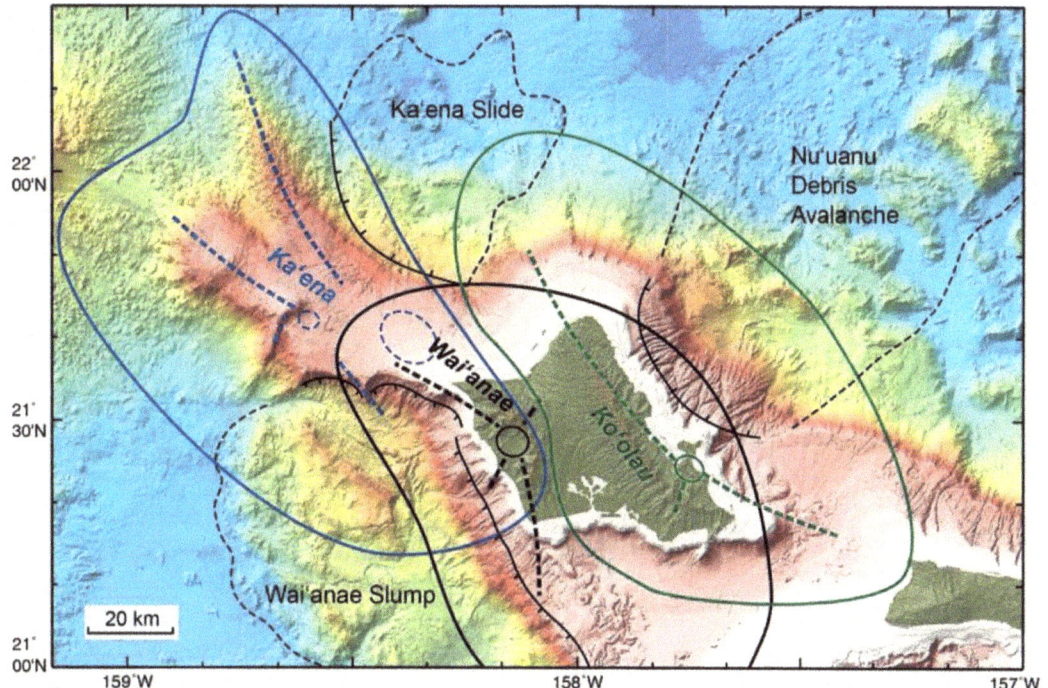

This map shows the distribution of the three volcanoes – the Ka'ena, Wai'anae and Ko'olau – now thought to have made up the region of O'ahu. Bold dashed lines delineate possible rift zones of the three volcanoes. Also shown are the major landslide deposits around O'ahu.

Kaua'i

Kaua'i is considered by secular geologists to be the oldest of the main Hawaiian Islands. With an area of 562 square miles, it is the fourth largest of these islands and the 21st largest island in the United States. Known also as the Garden Isle, Kaua'i lies 105 miles northwest of O'ahu. This island is the site of Waimea Canyon State Park. Kaua'i is a shield volcano.

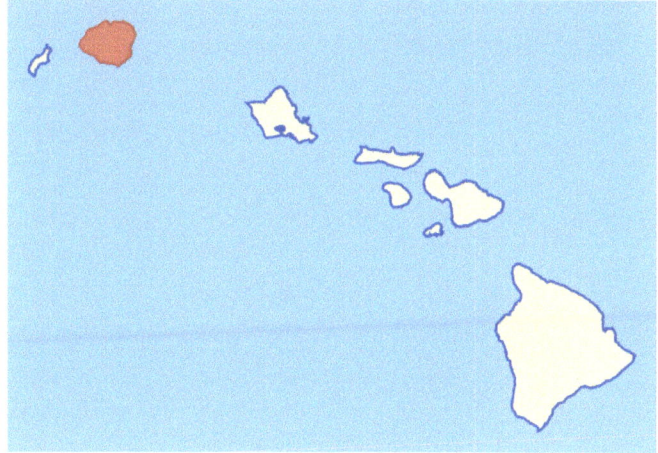

The island of Kaua'i

Satellite view of Kauaʻi

Niʻihau

Niʻihau is the westernmost and seventh largest inhabited island in Hawaiʻi. It is 17.5 miles southwest of Kauaʻi across the Kaulakahi Channel. It has an area of 69 square miles. It is a shield volcano.

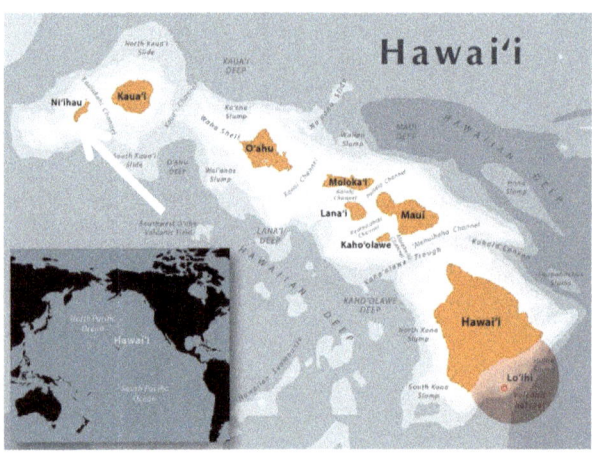

Aerial view of Niʻihau Island

What about Continental Drift, Plate Tectonics and Hot Spots?
In the 1950s as the ocean floor began to be mapped, seamounts, and particularly the Hawaiian Islands, were a mystery, along with the Yellowstone area. What do these two very different areas have to do with each other? We will get to that in just a bit. First, let's take a look at Continental Drift.

Continental Drift was initially proposed in the 1800s to attempt to explain the presence of fossils in isolated and water locked places like Australia and Antarctica. The belief was that if all living things had a common origin, the continents must have been one continent at one time. So how do you explain finding fossils of similar kinds on different continents? Well, some geologists believe the continents must have drifted away from each other at some time in the past after the species were well developed. That idea has evolved through the years, culminating in modern plate tectonics.

> **Continental Drift and the Bible**
>
> There is another way to explain what some say appears to be Continental Drift. As opposed to moving continents, the Flood, with its violent, turbulent waters, would have torn up the one land mass that was part of the original creation and left land masses that we now call continents, and basins that now hold the oceans of the world.

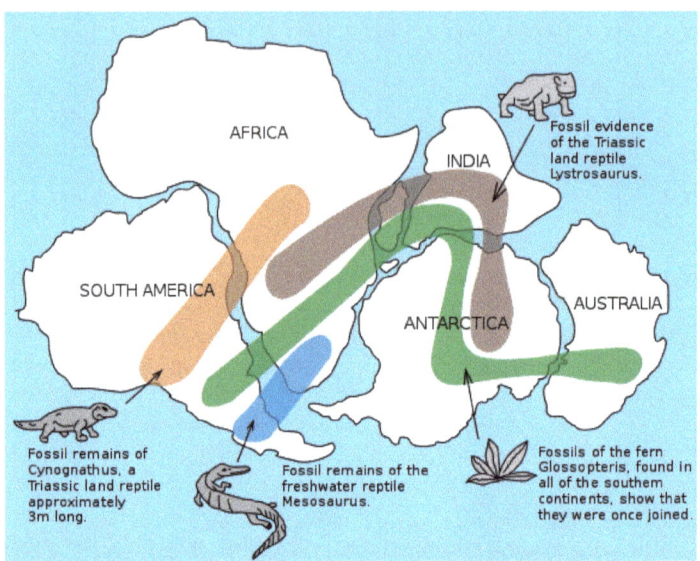

The distribution of fossils across the continents is what some scientists point to as evidence of the existence of Pangaea

Plate tectonics is an idea that stresses that the earth's crust is made of independent plates that move in relation to each other. According to secular scientists, the earth began 4.6 billion years ago. Since that time, the one original continent has broken up and moved around the globe at least twice. Over the last 200 million years, Pangea has supposedly been drifting apart into defined continents and will continue to drift until they come together again in the future into another united land mass. Some imagined that plate tectonics was able to explain why we have more volcanism in places like the west coast of the US, which is supposed to be located on a plate where two pieces of the crust of the earth are moving against each other. But this presented a mystery for Hawai'i and Yellowstone, because they did not fit this model. They have nothing to do with plate boundaries, which is where most of the volcanism in our world seems to be found.

Hence, they needed a different explanation for the creation of those two areas, and areas like them. Enter *Hot Spots*!

Hot Spots – Many geologists call places in the earth where lava is erupting, *hot spots*. A hot spot is an anomaly in modern geology. Some geologists believe that hot spots are connected to a magma chamber below the earth's surface. The hot spot idea basically says that as the crust shifts, the volcanic area in question is formed from lava rising from deep under the earth's crust to extrude out on to the earth's surface through a weak spot in the crust. That portion of the crust slides over the Hot Spot, where eruptions of lava find an opening. The Hawaiian Islands supposedly formed as the Plate on which they rest has slid over the stationary Hot Spot. It certainly is an idea. But has it ever been observed? No. Can it be scientifically tested and repeated? No. Therefore, the Hot Spot idea 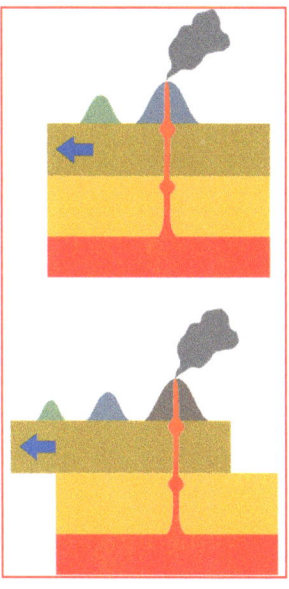 fits better in a different category of study, rather than science. It fits better in a philosophy department the same way the Genesis Flood fits. These are ideas, not science. The ideas surrounding hot spots and plate tectonics are a very hotly debated topic among geologists, regardless of their worldview. Time and more data will be necessary before definitive conclusions can be made. But such is the world of geology!

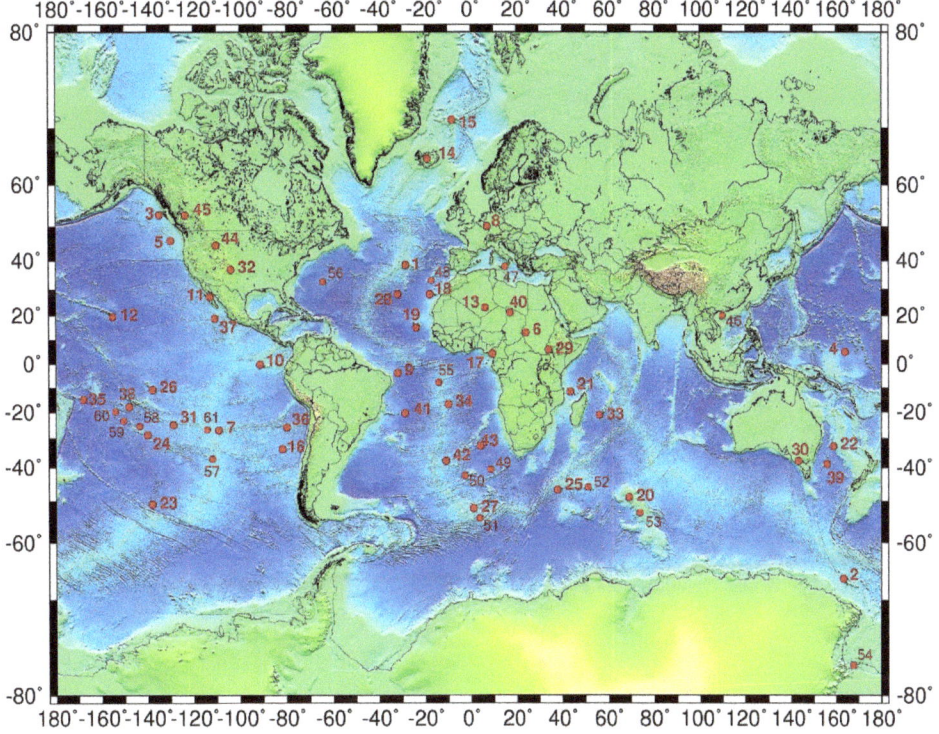

Map identifying major volcanic areas of the world, known as hot spots

Earthquakes on the Big Island of Hawai'i

When we think of volcanism on the Big Island, we usually think of erupting lava, not earthquakes. But where there is volcanism, there are earthquakes. Thousands of earthquakes occur on the Big Island of Hawai'i every year. These are thought to be the result of the prevalent present volcanism of the Big Island. It is thought that the active movement of the magma beneath the surface of the earth is what triggers the earthquakes. The USGS states, "Commonly, earthquakes occur ahead of intruding magma, permitting us to anticipate the probable location of the outbreak of lava at the surface." Words commonly associated with earthquakes and volcanism are ***tectonic earthquakes*** (dealing with faults and fault zones) and ***volcano earthquakes*** (dealing directly with the movement of magma). Both seem to be influenced by the behavior of magma underneath the earth's surface. The two most destructive earthquakes in Big Island recorded history were recorded in 1868, in which 81 people were killed, and in 1975 along the southeast coast.

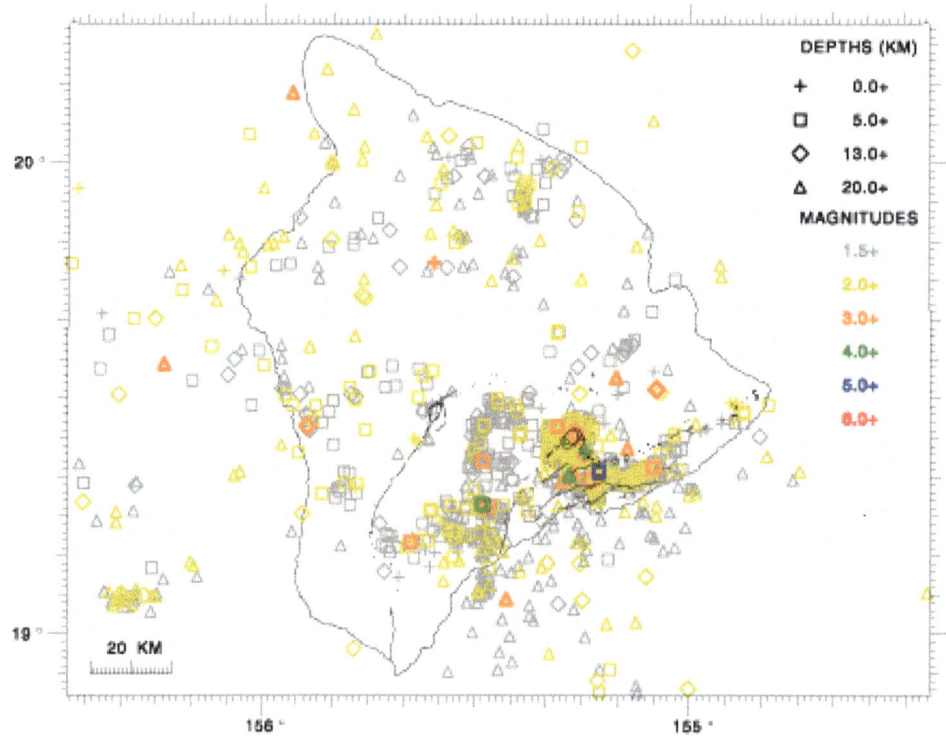

Earthquake activity on and around the Big Island of Hawai'i for the Year 2000

In 1960, rifts, earthquakes and lava flows from the Kapoho Crater, part of the Kilauea Caldera, were responsible for causing the destruction to the village of Kapoho, located to the east of the Kapoho Crater.

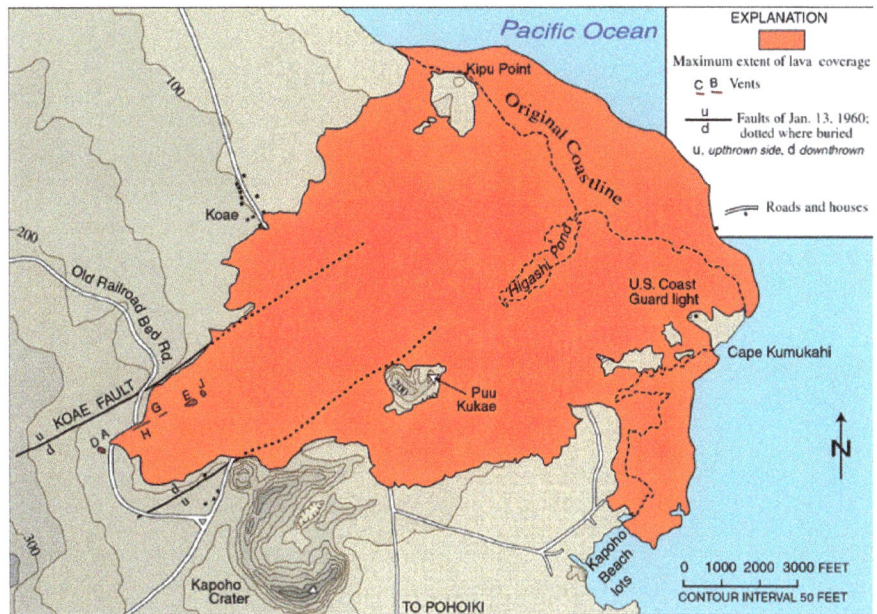

Map of the lava flow from the Kapoho crater (1960, part of the Mauna Loa complex of calderas, vents and craters) just to the southwest of the flow. The map also shows the two major fault zones which were the source of the major earthquake of 1975.

A fault in Kapoho village produced from the Kapoho eruption on January 13, 1960
Photo courtesy of the USGS

Canyons, ravines, and lava tubes of the Hawaiian Islands

One of the prominent features of the Hawaiian Islands are the deeply eroded volcanic mountains. These *V-shaped* ravines have been cut by water. V-cut canyons or ravines are recognized as having been cut rapidly by water. But it is

99

not simple water-run-off from years of rain. The prominent mountains of the Hawaiian Islands are made of very hard basalt lava. It is highly unlikely that normal rain would have made much of an impact on these hard basalt mountains. In fact, the word *basalt* means *hard*. It is one of the hardest rocks I have broken. Geologists date these v-shaped canyons and ravines as millions of years old. As no one saw what transpired when these mountains were formed, using our Biblical framework, they cannot be that old. Working within the Biblical framework, a reasonable idea would be that these mountains were carved by water that covered them during or shortly after formation, most likely during the receding stage of the Flood about 4,500 years ago.

The volcanic mountains of Oʻahu show a tremendous amount of erosion; were these ravines cut slowly and over millions of years, or were they cut rapidly and with a high velocity of water? The Basic Biblical Framework would dictate that these mountains were cut rapidly and with a high velocity of water.

The next picture is from the Big Island. The rounded basalt boulders indicate that a high velocity of water current was involved in rapidly eroding and transporting them. High velocity, not gradual erosion, is what is required to create the rounded boulders from hard basalt.

Water-tumbled basalt boulders, Hawai'i

Lava tubes are formed when the outer crust of a lava flow cools and hardens faster than the inside of the flow, leaving a hollow tube when finished.

 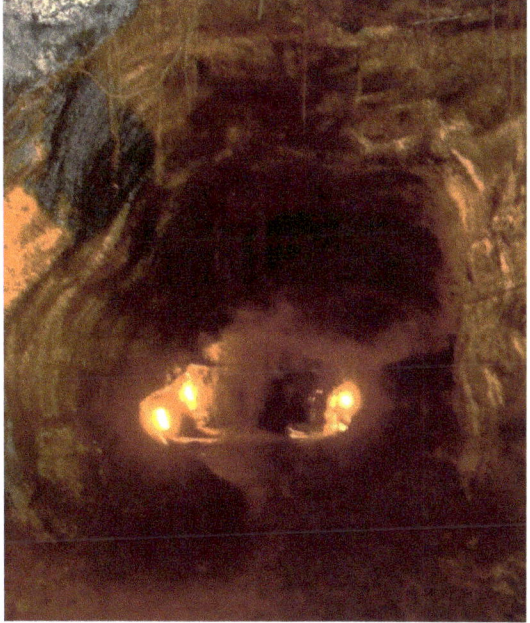

Inside a lava tube, the Big Island

Hawai'i and the Ice Age

Another very interesting feature on the Big Island of Hawai'i is located on Mauna Kea.

Glacial evidence on Mauna Kea, outlining terminal moraines.

What? Glacial evidence on the Big Island? According to the United States Geological Survey, there is evidence of glaciers that once covered parts of Mauna Kea. Now, these glaciers are not just mounds of snow like what falls on the summit of Mauna Loa today. Glaciers form where the accumulation of snow exceeds the melting and sublimation over many years, often centuries. Secular geologists teach that glaciers are persistent bodies of dense ice that are constantly moving under their own weight. Consequently, glaciers slowly deform and flow due to stresses induced by their weight, creating crevasses and seracs (blocks or columns of glacial ice, often formed by intersecting crevasses on a glacier). Commonly house-sized or larger, they are dangerous to mountaineers since they may topple with little warning.

The previous photo of Mauna Kea shows the evidence for glacial moraines. A moraine is any glacially formed accumulation of unconsolidated glacial debris (soil and rock) that occurs in currently glaciated and formerly glaciated regions on Earth. These are expected in places like Yellowstone and Washington State, where glaciers were active in the past. But Hawai'i? Are you kidding me? How could this be?

During the end stages of the Flood, there is good evidence that volcanism continued unabated for a few hundred years. The amount of ash erupted into the atmosphere would have blocked the sun's reflective energy making the atmosphere colder. The water from the oceans, having recently been infused with hot magma from the breaking up of the fountains of the great deep, would have warmed the ocean water slightly, increasing evaporation. Warm, moist air condensing in a cooler atmosphere would have been a perfect recipe for a type of major ice event, which geologists call an *Ice Age*.

Ice would have accumulated significantly in the northern hemisphere and in higher latitudes, including the young and fairly recent volcano, Mauna Kea on the north side of the Big Island. Water from the Flood was tied up in the worldwide glaciers for several centuries, lowering the ocean levels by several hundred feet. Later, as volcanic eruptions settled down, the sun's reflective energy would have returned and the ice sheets and glaciers would have begun to melt. Sea levels would have once again risen, producing flooding. Subsequent flooding could have caused some of the features of many of the Hawaiian Islands, including ravines and rounded boulders. The glacier(s) that had occupied the summit of Mauna Kea then disappeared. Today we have snow at the higher elevations of the volcanoes on the Big Island, but no more glaciers, just the evidence that they had once been there.

Hawai'i is vibrant and geologically alive. More than just an exotic vacation spot, the Islands are a laboratory for the study of the effects of the Genesis Flood. They shout of a catastrophic origin, and all within the same recent time frame – the Genesis Flood and its immediate aftermath. Through the Islands, the Lord has given us a fantastic window into His workings. I hope you have enjoyed your study!

Thought Questions

1. What kinds of evidence support a young age for the volcanoes of Hawai'i?

2. What is the predominant type of volcano making up the Hawaiian Islands?

3. Describe the rock *basalt*.

4. Describe the rock *tuff*.

5. Describe the rock *palagonite*.

6. What are cinder cones? What kind of rock is formed as a result of a cinder cone eruption?

7. How did Diamond Head get its name?

8. What is the Biblical view of the deeply cut ravines on the mountains of the Hawaiian Islands?

9. What are the dates, which are given in millions of years, for the Hawaiian Islands based on?

10. Describe the reason why we might find glacial moraines on the summit of Mauna Kea.

Activity: Using the Basic Biblical Framework, describe and place the volcanism of the Hawaiian Islands.

Appendix A
Comprehensive Exam

1. The name for the rock that is made of welded ash and bits of volcanic rock is:
   a) Tachylite
   b) Tuff
   c) Ash
   d) Pele's Tears
2. The word *mafic* stands for:
   a) Magma and quartz
   b) Magma and potassium feldspar
   c) Magnesium-like
   d) Magnesium and iron
3. Palagonite is:
   a) Altered rhyolite
   b) Altered opal
   c) Altered dacite
   d) Altered basalt
4. A phreatic eruption involves:
   a) Hot magma and steam
   b) Hot water and steam
   c) Hot lava and steam
   d) Hot rocks and ash
5. The Basic Biblical Framework involves:
   a) Two stages of the Flood
   b) Three stages of the Flood
   c) Six stages of the Flood
   d) The Flood and the Ice Age
6. In order for radiometric dating to work, what must be accepted?
   a) The use of plutonic rocks
   b) The use of metamorphic rocks
   c) A set of assumptions
   d) A set of volcanic rocks
7. An isotope is:
   a) An ordinary element
   b) A large element
   c) A varied form of the original element
   d) A new element

8. An atom is made up of three parts:
   a) Plutons, electrons and neutrons
   b) Protons, electrons and neutrons
   c) Pulsars, electrons and neutrons
   d) Particles, elements and neutrons
9. Hawai'i has how many national parks?
   a) Two
   b) Three
   c) Four
   d) Ten
10. The main difference between a shield volcano and a stratovolcano is:
    a) Shape
    b) Size
    c) Ice
    d) Depth
11. The main difference between basalt and rhyolite is:
    a) Color
    b) Amount of quartz
    c) Amount of feldspar
    d) Amount of vesicles
12. Olivine is:
    a) A rock type
    b) An element
    c) A rock-forming mineral
    d) An altered rock
13. Basalt typically erupts as:
    a) Hot pink lava
    b) Hot gray lava
    c) Cool rhyolite lava
    d) Hot black lava
14. A cinder cone typically erupts:
    a) Quiet lava
    b) Quiet andesite
    c) Showers of basalt
    d) Hot steam
15. The temperature of erupting basalt can be as high as:
    a) 500^0
    b) 700^0
    c) 250^0
    d) 2000^0

16. Pahoehoe is a type of:
   a) Pumice
   b) Sharp, clinker lava
   c) Ropy lava
   d) Vesicular lava
17. The most massive shield volcano on Earth is:
   a) Kilauea
   b) Kohala
   c) Oʻahu
   d) Mauna Loa
18. A submarine volcano is also called a:
   a) Seamount
   b) Turbidity mount
   c) Sand dune
   d) Sea cone
19. The term used to describe the idea of shifting continents is:
   a) Glacial drift
   b) Continental drift
   c) Plate Tectonics
   d) Mountain Tectonics
20. An example of a caldera/shield volcano is:
   a) Diamond Head
   b) Kilauea
   c) Koko Bay
   d) Maui

Answers to the Comprehensive Exam

1. b
2. d
3. d
4. d
5. a
6. c
7. c
8. b
9. a
10. a
11. b
12. c
13. d
14. c
15. d
16. c
17. d
18. a
19. c
20. b

Picture Credits

Lesson One:
Biblical Timeline Diagram credit: Patrick J. Nurre, 8; Genesis Flood diagram: 1st Photo: NOAA 2nd picture: URL: http://geomaps.wr.usgs.gov/parks/noca/nocageol4c.html; 3rd picture: public domain; 4th picture: Photo: Marli Miller, USGS; 5th picture: USGS photo, diagram by Patrick and Vicki Nurre, 10; Geological events of the Genesis Flood: Diagram by Patrick Nurre 11;

Lesson Two:
The Hawaiian Islands: By United States Geological Survey (USGS) - http://hvo.wr.usgs.gov/volcanoes/, Public Domain, https://commons.wikimedia.org/w/index.php?curid=196173, 13; The Periodic Table: By Sandbh - Own work, CC BY-SA 4.0, https://commons.wikimedia.org/w/index.php?curid=53697362, 15; Configuration of Carbon-12 and Carbon-14, Diagram by Vicki Nurre, 16; Radioactive atom: Source: NRC at http://www.nrc.gov/reading-rm/basic-ref/glossary/full-text.html, 17; Parent-daughter decay process: Diagram by Vicki Nurre, 18; Uraium radioactive decay, Chart by Vicki Nurre, 19; Modern Geologic Column: http://www.bing.com/images/search?pq=geologic+time+&sc=814&sp=2&sk=IA1&q=geologic+time+scale+chart&qft=+filterui:license L2_L3&FORM=R5IR42#view=detail&id=0548D46FA426B2733D713B5A16EE55BFDB889547&selectedIndex=0, 20; Potassium-argon 'ages' for whole rock and mineral concentrate samples from the lava dome at Mount St Helens that was in reality about 10 years old, Chart by Patrick Nurre, 22; Mt. St. Helens lava dome: Chart courtesy of the USGS, 22; Radiometric discrepancies, Mt. St. Helens:: Chart by Patrick Nurre, 23; Other radiometric discrepancies: Chart by Patrick Nurre, 23; Radiometric discrepancies, Hawaii: Chart by Patrick Nurre, 24; Radiometric discrepancies, Grand Canyon: Photo by Patrick Nurre, 24; Mt. Ngaruohe: by © Guillaume Piolle /, CC BY 3.0, https://commons.wikimedia.org/w/index.php?curid=31740809, 25; Lava flow: Photo by Vicki Nurre, 26;

Lesson Three:
The Basic Biblical Framework: Diagram credit: Patrick Nurre, 28; Diagram: Volcanic hazards Usgs credit, 29; Acid rain effect: CreditBy Slick - Own work, CC0, https://commons.wikimedia.org/w/index.php?curid=22257663, 30; pH scale: Diagram by Vicki Nurre, 30; Lahar: Picture creditBy Robin Holcomb, U.S. Geological Survey - Archived source link, Public Domain, https://commons.wikimedia.org/w/index.php?curid=5883394, 30; Lahar: By Jeffrey Marso, USGS geologist - source, small version with description here, Public Domain, https://commons.wikimedia.org/w/index.php?curid=5848803, 30; Teleoceras fossil: CC BY-SA 2.0, https://commons.wikimedia.org/w/index.php?curid=126389, 31; Trees at Mt. St. Helens: USGS by Lyn Topinka, 31; Ash cloud and pyroclastic flow: By C.G. Newhall - http://volcanoes.usgs.gov/Imgs/Jpg/Mayon/32923351-020_caption.html, Public Domain, https://commons.wikimedia.org/w/index.php?curid=246427, 31; Lava flow: Jon Rosenberg, 32; Lava flow: Photo courtesy of the USGS, 32; Effects of volcanoes: CreditBy cflm (talk) - Own work. Derived from File:Volcanic injection.jpg by SEWilco(en.wp), released under PD-USGov-Interior-USGS. Based on data from the US Geological Survey. Coloured using Inkscape., Public Domain, https://commons.wikimedia.org/w/index.php?curid=7817020, 32; Damage at Mt. St. Helens: Four photos courtesy of USGS by Lyn Topinka, 33; Mt. St. Helens, pre-eruption: Photo by Harry Glicker, courtesy USGS, 33; Mt. St. Helens post-eruption: Photo by Harry Glicken, courtesy of USGS, 33; Mt. St. Helens after 20 years: Photo by Gene Iwatsubo, courtesy of USGS, 33;

Lesson Four
Fissure volcano on Hawai'i: Public Domain, https://commons.wikimedia.org/w/index.php?curid=17445, 38; Shield volcano, Mauna Kea: By Nula666 - Own work, CC BY-SA 3.0, https://commons.wikimedia.org/w/index.php?curid=11558567, 38; Dome volcano, Mt. St. Helens: Photo by Steve Schilling, USGS, 38: Cinder cones: Photo by Vicki Nurre 38; Composite or Stratovolcano, Mt. Fugi: By hogeasdf - http://www.flickr.com/photos/9177053@N05/4071658790/, CC BY 3.0, https://commons.wikimedia.org/w/index.php?curid=9100465, 38; Caldera volcano, Crater Lake: By Zainubrazvi - Own work, CC BY-SA 3.0, https://commons.wikimedia.org/w/index.php?curid=2033872, 38; Types of volcanoes: Courtesy of USGS, 39; Fissure volcano: Courtesy of USGS, 39; Mauna Loa: By J.D. Griggs - U.S. Geological Survey, Public Domain, https://commons.wikimedia.org/w/index.php?curid=835932, 40; Chaiten Volcano: Photo by By Sam Beebe - Volcán Chaitén, CC BY-SA 2.0, https://commons.wikimedia.org/w/index.php?curid=6495901, 40; Dome at Mt. St. Helens: Photo courtesy of USGS, 41; Dome growth: Diagram courtesy of USGS 41; SP Crater, Arizona: Photo courtesy of USGS, http://pubs.usgs.gov/fs/2001/fs017-01/, Public Domain, https://commons.wikimedia.org/w/index.php?curid=1260318, 42; Mauna Kea: Photo by Kelly E. Fast, Courtesy of NASA, 42; Mt. Fuji: Photo by DoctorJoeE - Own work, CC BY-SA 4.0, https://commons.wikimedia.org/w/index.php?curid=46646708 Mt. Fuji, Japan, 43; Mt. Hood, Oregon: Photo by Oregon's Mt. Hood Territory. - http://www.fhwa.dot.gov/byways/photos/62736., Public Domain, https://commons.wikimedia.org/w/index.php?curid=715348, 43; Mt. St. Helens: Photo courtesy of USGS. 44; Shield volcano v. composite volcano: Diagram courtesy of USGS, 44; Yellowstone map: Courtesy of USGS, 45;
Laccolith diagram: By en:User:Erimus, User:Stannered - en:Image:Laccolith.JPG, Public Domain, https://commons.wikimedia.org/w/index.php?curid=3713849 diagram of a laccolith, 46; Laccolith: By Jstuby at en.wikipedia - Own workTransferred from en.wikipedia, Public Domain, https://commons.wikimedia.org/w/index.php?curid=17781236, 47; Laccolith: Photo by Patrick Nurre, 47; Square Butte: Picture credit By Mike Cline - Own work, CC BY-SA 4.0, https://commons.wikimedia.org/w/index.php?curid=40798388, 48; Devil's Tower:Public Domain, https://commons.wikimedia.org/w/index.php?curid=615805, 48; La Sal Mountains: Photo by Patrick Nurre, 49; La Sal Mountains: Photo by Patrick Nurre, 49; Ring of Fire Map: By Gringer (talk) 23:52, 10 February 2009 (UTC) - vector data from [1], Public Domain, https://commons.wikimedia.org/w/index.php?curid=5919729, 50; Rifts around the world: Courtesy of USGS found at https://pubs.usgs.gov/gip/dynamic/baseball.html, 51; *Parts of a Volcano: By MesserWoland - own work created in Inkscape, CC BY-SA 3.0,* https://commons.wikimedia.org/w/index.php?curid=1266825, *51;* Volcano cross-section: Free use;

Lesson Five
Hawaiian Eruption: By © Sémhur / Wikimedia Commons, FAL, https://commons.wikimedia.org/w/index.php?curid=2874732,55; Pahoehoe lava: Photo by Jon Rosenberg, 55; Strombolian Eruption and Mt. Stromboli, Italy; Credit By © Sémhur / Wikimedia Commons, FAL, https://commons.wikimedia.org/w/index.php?curid=2874911, 55; Stromboli: Photo by Wolfgangbeyer at the German language Wikipedia, CC BY-SA 3.0, https://commons.wikimedia.org/w/index.php?curid=34243, 55; Volcanian Eruption: By Sémhur (talk) - Own work, from the document about volcanism on the Portail sur la prévention des risques majeurs (web portal about the prevention of the

major risks) of the Ministère français de l'Ecologie, du Développement et de l'Aménagement durables (french Minister of the Ecology, Environment and Sustainable Development)., FAL, https://commons.wikimedia.org/w/index.php?curid=2875033, 56; Irazu Volcano: Photo by Rafael Golan - Own work, CC BY-SA 3.0, https://commons.wikimedia.org/w/index.php?curid=17990842, 56; Peléan eruption: By © Sémhur / Wikimedia Commons, FAL, https://commons.wikimedia.org/w/index.php?curid=2874781, 56; Mayon Volcano: Photo by C.G. Newhall - http://volcanoes.usgs.gov/Imgs/Jpg/Mayon/32923351-020_caption.html, Public Domain, https://commons.wikimedia.org/w/index.php?curid=246427 , 56; Plinian eruption: By © Sémhur / Wikimedia Commons, FAL, https://commons.wikimedia.org/w/index.php?curid=2874892,57; Redoubt Volcano. Public Domain, https://commons.wikimedia.org/w/index.php?curid=5768911, 57; Surtseyan eruption: By © Sémhur / Wikimedia Commons, FAL, https://commons.wikimedia.org/w/index.php?curid=2875010, 57; Surtsey: Public Domain, https://commons.wikimedia.org/w/index.php?curid=748137, 57; Submarine eruption: credit By © Sémhur / Wikimedia Commons, FAL, https://commons.wikimedia.org/w/index.php?curid=2874990,58; Loʻihi: John Smith and Brooks Bays - (Original text: "National Science Foundation / National Oceanic and Atmospheric Administration / The Hawaii Undersea Research Laboratory (HURL) http://www.soest.hawaii.edu/HURL/images/loihi_3d.gif http://www.soest.hawaii.edu/HURL/hurl_loihi.html), Public Domain, https://commons.wikimedia.org/w/index.php?curid=8266876, 58; Phreatic eruption: By Sémhur - own work / travail personnel.Inspired by the document about volcanism from the Portail sur la prévention des risques majeurs (web portal about the prevention of the major risks) of the Ministère français de l'Ecologie, du Développement et de l'Aménagement durables (french Minister of the Ecology, Environment and Sustainable Development)., CC BY-SA 3.0, https://commons.wikimedia.org/w/index.php?curid=2874846, 58; Mt. St. Helens: By USGS - United States Geological Survey - http://volcanoes.usgs.gov/Products/Pglossary/HydroVolcEruption.html, Public Domain, https://commons.wikimedia.org/w/index.php?curid=10016190, 58; Classification: The Comet program, courtesy of USGS found at http://www.goes-r.gov/users/comet/volcanic_ash/volcanism/navmenu.php_tab_1_page_5.0.0.htm, 59;

Lesson Six
Dark-colored rock-forming minerals: Photo by Vicki Nurre, 61; Light-colored rock-forming minerals: Photo by Vicki Nurre, 61; Yellowish orange iron oxide: Photo by http://www.photolib.noaa.gov/htmls/nur05020.htm. Transferred from en.wikipedia to Commons by PatríciaR., Public Domain, https://commons.wikimedia.org/w/index.php?curid=3432526, 62; Earth's most abundant minerals: Diagram by Vicki Nurre, 63; Basalt: Photo by Patrick Nurre, 63; Oxidized basalt: By http://www.photolib.noaa.gov/htmls/nur05020.htm. Transferred from en.wikipedia to Commons by PatríciaR., Public Domain, https://commons.wikimedia.org/w/index.php?curid=3432526, 63; Vesicular basalt: Photo by Patrick Nurre, 64; Pahoehoe lava: Photo by Patrick Nurre, 64; Basalt with olivine: Photo by Patrick Nurre 64; ʻAʻa basalt lava flow: Photo by Patrick Nurre, 64; Olivine: Photo by Patrick Nurre, 64; Papakolea Beach: Photo by Madereugeneandrew - Own work, CC BY-SA 4.0, https://commons.wikimedia.org/w/index.php?curid=40183783, 65; Aphanitic basalt: Photo by Patrick Nurre, 65; Basalt Porphyry: Photo by Patrick Nurre, 65; Scoria: Photo by Patrick Nurre, 66; Cinder: Photo by Doverbeach1 - Own work, Public Domain, https://commons.wikimedia.org/w/index.php?curid=11189012, 66; Cinders; Photo by Patrick Nurre, 66; Pele's Tears: By Jim D. Griggs, HVO (USGS) staff photographer[1][2] - http://volcanoes.usgs.gov/images/pglossary/PeleTears.php, Public Domain, https://commons.wikimedia.org/w/index.php?curid=700026, 66; Tachylite: Photo by Patrick Nurre, 67; Vitrophyre: Photo by Patrick Nurre, 67; Black sand: Photo by Patrick Nurre, 68; Punaluu beach: Photo by Vicki Nurre, 68; Ribbon bombs: Photos by Patrick Nurre. 68; Spindle bomb: found by Rob McConnell in the Mojave National Preserve, California, USA, Public Domain, 69;

Lesson Seven
The Emperor-Hawaiian Seamount Chain:By National Geophysical Data Center/USGS - http://www.ngdc.noaa.gov/mgg/image/2minrelief.html, Public Domain, https://commons.wikimedia.org/w/index.php?curid=617615, 71; Emperor-Hawaiian Seamount Chain: Public Domain, https://commons.wikimedia.org/w/index.php?curid=618807, 72; The Hawaiian Island Chain: by Jacques Descloitres - File:Hawaje.jpgOriginal source: NASA. Image courtesy Jacques Descloitres, MODIS Land Rapid Response Team at NASA GSFC. (IotD Date: 2003-06-03. IotD ID: 15304), Public Domain, https://commons.wikimedia.org/w/index.php?curid=11359528, 72; The Big Island of HawaiʻI map: By Mapmaunaloa.png: HawaiʻI Volcano Observatory, USGSderivative work: Richardprins (talk) - Mapmaunaloa.png, Public Domain, https://commons.wikimedia.org/w/index.php?curid=10852262, 73; Kohala: Public Domain, https://commons.wikimedia.org/w/index.php?curid=3745520, 73; Mauna Kea:
By Nula666 - Own work, CC BY-SA 3.0, https://commons.wikimedia.org/w/index.php?curid=11558567, 73;
Hualalai; Photograph by J. Kauahikaua. - originTransferred from en.wikipedia; transferred to Commons by User:Innotata using CommonsHelper., Public Domain, https://commons.wikimedia.org/w/index.php?curid=15042437, 74;
The Big Island of Hawaiʻi: by NordNordWest - own work, usingUnited States National Imagery and Mapping Agency dataU.S. Geological Survey (USGS) data, CC BY-SA 3.0 de, https://commons.wikimedia.org/w/index.php?curid=7015666, 74;
Mauna Loa: By J.D. Griggs - U.S. Geological Survey, Public Domain, https://commons.wikimedia.org/w/index.php?curid=835932, 75; Mauna Loa; Courtesy of the USGS, 75; Fissure eruption: Courtesy of the USGS, 76; Map of Mauna Loa and Kilauea: Courtesy of the USGS, 76; Kilauea: By Puu_Oo_looking_up_Kilauea.jpg: USGSderivative work: Avenue (talk) - Puu_Oo_looking_up_Kilauea.jpg, Public Domain, https://commons.wikimedia.org/w/index.php?curid=17010236, 77; Kalapanna: By J.D. Griggs - U.S. Geological Survey, Public Domain, https://commons.wikimedia.org/w/index.php?curid=835932, 77; Puʻu ʻOʻo creater: Courtesy ofUSGS, 78; Puʻu ʻOʻo eruption: Photo by J.D. Griggs on February 4, 1985 courtesy of the USGS, 79; Puʻu ʻOʻo map: USGS, 79; Satellite image: PuʻuʻOʻo Picture courtesy of the USGS, 80; Pahoa, Lava from Puu Oo cinder cone: By Madereugeneandrew - Own work, CC BY-SA 4.0, https://commons.wikimedia.org/w/index.php?curid=39413714, 80; Lava on road: Courtesy of USGS, 81; PuʻuʻŌʻō: Photo by J.D. Griggs - USGS HVO, Public Domain, https://commons.wikimedia.org/w/index.php?curid=925189, 81; Pāhoa: Photo by Madereugeneandrew - Own work, CC BY-SA 4.0, https://commons.wikimedia.org/w/index.php?curid=39415332, 81; Loʻihi:
By Hawaii_Island_topographic_map-en.svg: *Hawaii_Island_topographic_map-fr.svg: Sémhurderivative work: Kmusser (talk)derivative work: Kmusser (talk) - Hawaii_Island_topographic_map-en.svg, CC BY-SA 3.0, https://commons.wikimedia.org/w/index.php?curid=6085217, 82; Haleakala: By Karte: NordNordWest, Lizenz: Creative Commons by-sa-3.0 de, CC BY-SA 3.0 de, https://commons.wikimedia.org/w/index.php?curid=27552397, 83; Haleakala crater:
By Navin75 - Haleakala NPUploaded by Hike395, CC BY-SA 2.0, https://commons.wikimedia.org/w/index.php?curid=24963865, 83;
Maui Komohana By Sara Golemon - Another view of Puʻu KuʻkuiUploaded by hike395, CC BY-SA 2.0, https://commons.wikimedia.org/w/index.php?curid=6449760 Map: Haleakalā National Park

By National Park Service, Harpers Ferry Center - http://www.nps.gov/carto, Public Domain, https://commons.wikimedia.org/w/index.php?curid=9688873, 83;
Maui deposits: Courtesy of the USGS, 84; Kaho'olawe: By MattWright - Based on the SVG file that was released to the public domain by David Benbennick at, Public Domain, https://commons.wikimedia.org/w/index.php?curid=2484385, 86; Kahoolawe: By Landsat satellite image from NASA - http://coralreefs.wr.usgs.gov/kahoolawe.html, Public Domain, https://commons.wikimedia.org/w/index.php?curid=1982841, 86; Lana'i:
By MattWright - Based on the SVG file that was released to the public domain by David Benbennick at Wikipedia: Image:Map_of_Hawaii_highlighting_Kalawao_County.svg, Public Domain, https://commons.wikimedia.org/w/index.php?curid=2484406, 86; Lanai: By Landsat satellite image from NASA - http://coralreefs.wr.usgs.gov/lanai.html, Public Domain, https://commons.wikimedia.org/w/index.php?curid=1982844, 87; Moloka'i: By MattWright - Based on the SVG file that was released to the public domain by David Benbennick at Wikipedia: Image: Map_of_Hawaii_highlighting_Kalawao_County.svg, Public Domain, https://commons.wikimedia.org/w/index.php?curid=2484422, 87; Eastern Moloka'i; By Travis.Thurston - Own work, CC BY-SA 3.0, https://commons.wikimedia.org/w/index.php?curid=16714290, 88; Moloka'i; By Forest & Kim Starr, CC BY 3.0, https://commons.wikimedia.org/w/index.php?curid=6192687, 88; Bathymetric image:
By USGS: Barry W. Eakins, Joel E. Robinson,Japan Marine Science and Technology Center: Toshiya Kanamatsu, Jiro Naka,University of Hawai'i: John R. Smith,Tokyo Institute of Technology: Eiichi Takahashi, andMonterey Bay Aquarium Research Institute: David A. Clague - This bathymetry map (PDF) from this USGS page, Public Domain, https://commons.wikimedia.org/w/index.php?curid=377937, 89; O'ahu: By Karte: NordNordWest, Lizenz: Creative Commons by-sa-3.0 de, CC BY-SA 3.0 de, https://commons.wikimedia.org/w/index.php?curid=27552397, 90; Ko'olau Range: Photo by Commander Grady Tuell, NOAA Corps - Archived source link, Public Domain, https://commons.wikimedia.org/w/index.php?curid=16533399, 90; Palagonite:
Photo by B.navez - Own work, CC BY-SA 3.0, https://commons.wikimedia.org/w/index.php?curid=10568000, 91; Diamond Head: Photo by Brian Snelson - originally posted to Flickr as Diamond Head Crater, Oahu, CC BY 2.0, https://commons.wikimedia.org/w/index.php?curid=12228086, 91; Koko Bay Crater: Photo by No machine-readable author provided. Mbz1 assumed (based on copyright claims). - No machine-readable source provided. Own work assumed (based on copyright claims)., CC BY-SA 3.0, https://commons.wikimedia.org/w/index.php?curid=1996752, 91; Wai'anae Range: Photo by Joel Bradshaw - Own work, Public Domain, https://commons.wikimedia.org/w/index.php?curid=8852721, 92; Oahu: By MattWright - Based on the SVG file that was released to the public domain by David Benbennick at Wikipedia: Image:Map_of_Hawaii_highlighting_Kalawao_County.svg, Public Domain, https://commons.wikimedia.org/w/index.php?curid=2484433, 92; Ka'ena, Wai'anae and Ko'olau:
Image credit: J. Sinton et al / University of Hawai'i's School of Ocean and Earth Science and Technology, 93;
Map: Niihau, creditBy Mapbliss - Own work, CC BY-SA 4.0, https://commons.wikimedia.org/w/index.php?curid=42857166, 93; Ni'ihau Island: By Polihale at English Wikipedia, CC BY-SA 3.0, https://commons.wikimedia.org/w/index.php?curid=3677529, 94; Kaua'i: by MattWright - Based on the SVG file that was released to the public domain by David Benbennick at Wikipedia: Image:Map_of_Hawaii_highlighting_Kalawao_County.svg, Public Domain, https://commons.wikimedia.org/w/index.php?curid=2484270, 94; Kaua'i: Public Domain, https://commons.wikimedia.org/w/index.php?curid=195442, 95; Fossils and pngea: By Osvaldocangaspadilla - Own work, Public Domain, https://commons.wikimedia.org/w/index.php?curid=11310183, 96;
Hot spot: By Los688 - myown work, Public Domain, https://commons.wikimedia.org/w/index.php?curid=4618531, 96; Hot Spots: By JorisvS - Own work, CC BY-SA 4.0, https://commons.wikimedia.org/w/index.php?curid=44476501, 97; Big Island earthquake map: Courtesy of USGS, 98; Lava flow: https://hvo.wr.usgs.gov/kilauea/history/1960Jan13/kapoho7.htmlUSGS, 99; Fault: Courtesay of USGS, 99; Volcanic Mountains:Photo by Patrick Nurre, 100; Basalt boulders: Photo by Patrick Nurre, 101; Lava tube: Photo by Vicki Nurre, 101; Lava tube: Photo by Vicki Nurre, 101; Mauna Kea: By USGS - http://hvo.wr.usgs.gov/volcanowatch/2007/images/mkea_glaciation.jpg, Public Domain, https://commons.wikimedia.org/w/index.php?curid=16043977, 102;

Index

A'a' 5, 61 64
alluvium 71, 84
andesite 5, 38, 46, 56
andesitic 54, 56, 65
archipelago 13, 72, 85
ash 31, 33, 34, 42, 45, 47, 51, 54, 55, 57, 58, 65, 79, 84, 90, 93, 102, 103
assumption (assume) 6, 7, 14, 17, 19, 21, 22, 24, 26
atoll, 13, 28, 71
atom 13, 15, 16, 17, 18, 26
atomic mass 13, 16
basalt porphyry 61, 65
basalt 5, 23, 24, 39, 40, 46, 61, 62, 63, 64, 65, 67, 68, 70, 79, 88, 90, 92, 100, 101
bathymetry 71, 89
caldera 38, 40, 45, 46, 54, 74, 76, 78, 98, 99
catastrophism (catastrophic) 6, 7, 8, 11, 35, 45, 54, 81, 103
caustic 28, 29
channelized erosion 6, 11, 47
chronology 6, 8, 11, 103
cinder 38, 42, 61, 65, 66, 79, 80, 82
cinder cone/volcano 38, 42, 79, 80, 82,
composite volcano 38, 41, 43, 44,
corruption, 28, 29
dacite 5, 13, 22, 56
deterioration 28, 29
dikes 62, 67
dome volcano 38, 40, 41
earthquake 49, 50 51, 71, 98, 99
element 13, 14, 15, 16, 17, 18, 19, 26, 27, 61, 62, 63
Enlightenment 14, 25
eruption 7, 23, 29, 30, 32, 33, 34, 35, 37, 39, 44, 45, 46, 49, 54, 55, 56, 57, 58, 59, 60, 65, 68, 70, 78, 79, 81, 90, 97, 99, 103
fissure volcano 38, 39
fissure 38, 39, 55, 76
framework 6, 7, 8, 11, 12, 28, 35, 58, 100
geologic column (scale, table) 13, 20, 21, 24, 25
half-life 13, 17, 18, 19
history 4, 6, 7 8, 11, 14, 28, 45, 56, 98
hot spot 71, 95, 96, 97
intrusive 62, 67
iron oxide 61, 63
isotope 13, 15, 24, 26
lahar 28, 30, 33
lapilli 54, 55, 56
lava 4, 13, 14, 22, 24, 25, 26, 32, 37, 39, 40, 42, 44, 45, 46, 51, 54, 55, 56, 57, 58, 59, 61, 62, 63, 64, 65, 66, 68, 72, 74, 75, 76, 77, 78, 79, 80, 81, 84, 88, 90, 97, 98, 99, 100, 101

mafic 61, 62
magma, magmatic 24, 38, 46, 51, 54, 55, 56, 57, 58, 97, 98, 103
naturalism, naturalistic 6, 7, 21
obsidian 5
oxidation 62, 64
pahoehoe 5, 55, 61, 64
palagonite 71, 90, 91
Pele's Tears 55, 61, 66, 79
Pele's hair 55, 66, 79
periodic table 13, 15
pH Scale 28, 29, 30
phenocryst 61, 65, 67
philosophy 6, 19, 21, 26, 28, 97
phreatic eruption 54, 58
phreatomagmatic eruption 54, 57
pyroclastic 28, 31, 45, 56, 57, 65, 79, 90
radioactive, radioactivity 13, 14, 15, 16, 17, 18, 19, 21, 22, 24, 25, 26, 28
radiometric dating 13, 14, 17, 19, 21, 24, 25, 26, 28
rhyolitic, rhyolite 46, 54, 56, 65
rift 38, 50, 77, 79, 94, 96, 97, 99
Ring of Fire 49, 50
rock-forming minerals 4, 5, 61, 62, 63
science 4, 6, 14, 17, 21, 26, 28, 48, 97
scoria 5, 38, 42, 65, 66, 79, 90
seamount 13, 58, 71, 72, 81, 82, 95
secular 4, 6, 7, 8, 12, 21, 28, 34, 50, 84, 94, 96, 97, 102
sheet erosion 6, 11, 46, 48, 58
shield volcano 38, 40, 42, 55, 58, 72, 73, 74, 75, 76, 77, 83 85, 86, 87, 89, 90, 92, 93, 94
stratovolcano 38, 41, 43, 44
tachylyte (tachylyte) 61, 66, 67, 90
tectonic earthquake 71, 98
tephra   54, 60, 68
tuff cone 71, 90
uniformitarianism 6, 7, 11, 12, 19, 21, 25, 26, 28
veins 62, 67
vesicular basalt 5, 61, 63, 64, 65
viscosity   38, 46, 64
vitreous 62, 67
vitrophyre 5, 61, 67
volcanic vent 38, 54, 65, 74
volcano earthquake 71, 98
volcano 13, 29, 34, 35, 38, 39, 40, 41, 42, 43, 44, 45, 46, 49, 50, 51, 53, 54, 55, 56, 57, 58, 59, 61, 66, 70, 71, 72, 73, 74, 75, 76, 77, 78, 81, 82, 83, 84, 85, 86, 87, 89, 90, 92, 93, 94, 98, 103
worldview 6, 7, 11, 48

Patrick Nurre is from the beautiful state of Montana where as a young boy, he spent many of his Saturdays rockhounding near the Big Horn River. This early interest led him to a lifelong study of the world of geology. His experience has included extensive field study  in the Pacific Northwest, the Midwest and Plains states, the Southwestern U.S, and Israel. Patrick conducts classes and seminars in Seattle, and speaks at numerous homeschool conventions on geology and our young earth. He leads a variety of geology field trips every year, including one to Yellowstone National Park, where he helps families discover the Biblical geology of Yellowstone. Patrick's business, Northwest Treasures, is devoted to producing fine geology specimen kits and curricula from a young earth perspective. Patrick and his wife Vicki have three children and two grandchildren. They live in the Seattle, Washington area.

If you would like to contact Patrick about speaking or field trips:
northwestexpedition@msn.com
For a list of speaking topics: NorthwestRockAndFossil.com
Other books by Patrick Nurre – these are all also available with sample rock, mineral, and fossil kits at NorthwestRockAndFossil.com.

Rocks and Minerals for Little Eyes (PreK-3)
Fossils and Dinosaurs for Little Eyes (PreK-3)
Volcanoes for Little Eyes (PreK-3)
Geology for Kids (3-8)
Rock Identification Made Easy (3-12)
Rock Identification Field Guide
Fossil Identification Made Easy (3-12)
Fossil Identification Field Guide
Bedrock Geology (high school)
Rocks and Minerals: The Stuff of the Earth (high school)
Volcanoes, Volcanic Rocks and Earthquakes (high school)
The Geology of Yellowstone – A Biblical Guide
Genesis Rock Solid – A Biblical View of Geology
Fossils, Dinosaurs and Cave Men (high school)

www.ingramcontent.com/pod-product-compliance
Lightning Source LLC
Chambersburg PA
CBHW061929290426
44113CB00024B/2853